GM Fuel Injection Systems: Diagnosis and Repair

David Cadden

GM Fuel Injection Systems: Diagnosis & Repair

ISBN 1-878613-17-0

102290-5-PB

Table of Contents

Section I

Section II

Table of Contents

Section III

Table of Contents

Section IV

Section V

Purpose of This Manual

General Motors fuel-injected engines are fast becoming the majority of the GM fleet available to the aftermarket service industry. Only a small percentage of the new GM passenger cars and light trucks are being delivered with carburetors. Yet, evidence would indicate that many service technicians are not properly diagnosing and repairing GM fuel-injected engines. GM has spent untold millions of dollars creating volume after volume of technical repair manuals to guide the technician through the maze of high-tech evolutionary changes on the newer vehicles. The GM troubleshooting charts do work. Many technicians have found them to be quite satisfactory. Many have not. A lot of technicians never attempt to follow the charts at all. They merely read the bottom boxes to determine which parts are involved.

The purpose of this manual is to provide the field technician with a simple, yet comprehensive guide to proven successful diagnostic procedures for GM fuel-injected engine control systems. A technician who is knowledgeable concerning how each circuit and device works, how the individual circuits make up the working subsystems, and how the subsystems comprise a complete electronic fuel-injection system, can more quickly and confidently diagnose the engine performance and driveability problems which occur on engines controlled with that system.

The key emphasis is how it works and how to troubleshoot and test each part of the system. Leading a technician through an unexplained maze of diagnostic charts without bothering to inform the technician why he is being required to make such tests, has produced a history of misdiagnostic failures. In the process, millions of dollars have been wasted in time, replacement parts costs, and worst of all, the confidence and credibility of the technician to his peers and to the vehicle owner. The dreaded "comeback" has been increasing steadily. Customers report a loss of confidence in the technician. These "comebacks" cost everyone involved in the failed diagnosis a great deal of money and time.

A comprehensive and thorough diagnostic routine is absolutely mandatory. It doesn't guarantee to eliminate "comebacks", but it has been proven to be the only cost-effective way to reduce them to an acceptable level. This manual is structured for the technician who desires to follow a disciplined, logical sequence diagnostic routine. The systems identifications, specifications, circuit descriptions and test procedures are the essence of the material covered.

Purpose of Manual

(cont)

Becoming a successful GM fuel-injection systems technician will require three basic things. Achieving one or two will not be sufficient. The requirements are:

1. Professional tools and testers

2. Professional training

3. Professional data sources

From now on, the successful technician will have to stay abreast of annual changes. Each year, as the systems continue to evolve towards greater sophistication, the annual changes will require new tools, new information, and new training on how to use them.

This manual explains the need for, and how to use, many of these special tools and testers. The training on how to use these tools and testers to diagnose and repair these systems is being offered all over the country by many different companies and in many different formats. The information offered in this manual can be a good first source of data for these GM EFI engines. Keep in mind that this manual is designed to supplement your technical library, not replace it. The easy-access format should make this manual the first book you grab when a GM EFI vehicle arrives.

In summary, following a good logical sequence diagnostic procedure using this manual as your guide will in most cases produce winning results. If you want to be led blindly through traditional trouble trees, then this manual is not for you. If you want to know how the thing works, and how to test it, and you want a good, easy-access quick reference source, you've got it in your hands!

Course Outline

I. Introduction

A. GM EFI Diagnosis
  1. Driveability Problem
    a. Change from normal performance
    b. Tune-up time? Maybe...
  2. Driveability Symptom
    a. Hesitation
    b. Hard Starting
    c. Misfire
    d. Stalling
    e. Miscellaneous
  3. Logical Starting Place

C. Manual designed to aid the technician
  1. Specs
  2. System evolution and application
  3. Management systems theory
  4. Input and output circuits
  5. Controller pin identifiers
  6. Common troubleshooting procedures
  7. GM ignition systems
  8. Technical service bulletins
  9. Common EFI trouble points
  10. Glossary and acronyms

II. Logical Diagnostic Procedures

A. Set up shop disciplines
  1. Identify the complaint(s)
  2. Visual inspection
  3. Identify the vehicle
  4. Test basics
  5. GM Diagnostic Circuit Check
    a. Retrieve codes
    b. Clear codes and retest
    c. Repair hard code failures
    d. Test management subsystems

Course Outline

II. Logical Diagnostic Procedure (Cont.)

- B. Diagnostic report system
  - 1. Compose system report
  - 2. Compose GM EFI report
  - 3. Customize for your shop
  - 4. Use as sales report
- C. Basic EFI principles
  - 1. GM fuel management
  - 2. GM timing management
  - 3. GM idle management
- D. GM ignition Systems
  - 1. HEI
  - 2. HEI-EST
  - 3. HEI-EST-ESC
  - 4. DIS

III. Troubleshooting EFI

- A. Use GM Diagnostic Circuit Check
  - 1. Manual Check engine light
  - 2. Scanner
  - 3. Clear codes - retest
  - 4. Field service mode
  - 5. Solenoid check
  - 6. Fuel back-up control
  - 7. State 8 vs. State 9
- B. Individual circuit diagnosis
  - 1. ECM
  - 2. Inputs
    - a. CTS
    - b. TPS
    - c. MAP
    - d. MAF
    - e. MAT
    - f. Switches
    - g. Miscellaneous

Course Outline

III. Troubleshooting EFI (Cont.)

3. Outputs
   a. Injector(s)
   b. IAC
   c. EST control
   d. Solenoids
   e. Miscellaneous

IV. GM ECMs

A. Common ECM pin assignments
   1. Type 1 through type 6
B. Ascending year models have fault code variations
C. Testing circuits
   1. Test probe interface
   2. Scanner circuit readings

V. Common GM EFI Vehicle Tests & TSBs

A. Fuel delivery
   1. Pressure
   2. Flow
   3. Quality
B. Injector Cleaning
   1. Hook-up
   2. Procedures
C. Minimum idle adjustment
D. Injector balance test
E. Technical service bulletin tips

Table of Contents

Section I

Section I Outline

I. Introduction

A. Diagnosing GM fuel injection systems is easy.

1. First, let's define the question, "What is a driveability problem?"
    a. A variation in performance reflecting a change from normal operation.
    b. Many owners will deduce that this change signals the need for a diagnosis.
2. These changes in normal engine performance can be identified as specific problems or symptoms.
    a. Some vehicles will hesitate upon acceleration.
    b. Some engines will be hard to start.
    c. Sometimes the engine seems to misfire.
    d. Some might even stall or cut off.
    e. The list of possible symptoms can seem endless.
3. When any of these symptoms appear, it is time to decide where and how to logically detect the cause and effect a cure.
4. Trying to accomplish this based on only the knowledge of which symptom is present is not a conclusive diagnosis.
5. A professional EFI technician will have a developed, logical diagnostic sequence of measurements which will, in most cases, result in a conclusive diagnosis.

B. Diagnosing EFI engine performance problems requires several key items.

1. The right array of professional tools and testers.
2. A comprehensive library of up-to-date technical data.
3. The professional technician of today seeks out and acquires a comprehensive program of systems training.

C. This manual will aid the technician in this logical diagnostic procedure and provide pertinent systems information to help him accomplish his task.

1. The systems specifications are listed in Section I.
2. Section I describes each GM EFI system, its design and when released.
3. Each management system and subsystem is explored.
4. The input and output circuits are examined in simple yet detailed format.
5. The common engine controller connector pin identifiers will identify where each circuit terminates at the ECM.
6. The common troubleshooting procedures which are often used to cure GM driveability problems.
7. Additional information on troubleshooting and identifying each GM EFI system.
8. Technical service bulletins pertaining to GM EFI fixes are also presented.
9. Common fuel injection trouble points and how to correct them are covered.
10. The glossary and an acronym listing in the back of the manual will make it easier to understand the new terminology.

How To Use This Manual

This manual is set up into five easy-access sections designed to aid the technician in driveability diagnosis. The overall purpose of the manual is to allow the technician to use this manual during a driveability diagnosis for quick reference and instruction on system and circuit theory as well as troubleshooting information. Following the suggested diagnostic logical sequence found in Section II will, in most cases, result in a conclusive diagnosis.

After identifying the customer complaint(s), use the information found in Section I to carefully identify the vehicle and its systems. Also in Section I is a matrix layout called GM SYSTEM ID which can be used for this purpose. It is very important for the technician to identify the system accurately in order to direct his attention to the right subsystem controls.

While performing the Basic Engine Systems Tests, the technician can refer to Section II for additional information on the theory of operation and the troubleshooting procedures applicable to each type of GM ignition system. Many driveability complaints on GM vehicles revolve around ignition system malfunction.

Entering GM's self–diagnostic system using the Diagnostic Circuit Test is then explained in Section III. The proper procedure for retrieving and clearing GM fault codes is discussed. The General Fault Code Descriptions are also found in Section III. These definitions will generally inform the technician which circuit(s) need to be pinpoint tested. If a fault code has multiple definitions, the technician can find the Specific Fault Code Definitions listed by ascending year model in Section IV.

After the fault codes are clearly defined and the technician knows which circuit(s) are affected, the subsequent pages in Section III explain in detail the theory of operation and the circuit diagnosis of each input and output circuit used on GM EFI engines. Learn to quickly locate each device in the Section III Table of Contents.

Since GM EFI engines don't all use the same ECM, Section V contains ECM pin identifiers for each type of GM ECM used on EFI engines. These are classified as Type 1 through Type 6. This classification is for organizational simplicity only. The manufacturer does not classify or label them as such. This section will, in most cases, eliminate the need for the technician to refer to complete vehicle wiring diagrams. Since the circuit schematic is illustrated in Section III, and the circuit termination points are identified in Section IV, the technician can make tests at both ends of each circuit using these charts.

How To Use This Manual

Cont.

Section V contains testing procedures which are commonly used on GM fuel injected engines. Also, the technical service bulletins which have been issued concerning EFI driveability problems are condensed in this section and reformatted for easy referral.

Learning to use this manual is as easy as learning to access the information located in the master Table of Contents at the front of the manual or the individual Tables of Contents located at the beginning of each section.

This manual is not designed to replace your technical library, it is designed to supplement and enhance your library with information that the properly-trained EFI technician needs to work through a logical sequence diagnosis on GM fuel injected vehicles. Use it and enjoy it.

Warn Your Customers!

The advent of the microprocessor has brought about great changes to the automobile and the way we service them. Many automobiles sold today and in the last ten years have been equipped with many devices and systems which contain what is known as battery backed up RAM. In layman's terms, this means there are all kinds of gadgets on these vehicles such as programmable radios, trip computers, seats with memory, miles-to-empty computers, etc., which rely on the vehicle battery to keep certain volatile memory in storage.

It is absolutely necessary to forewarn your customers that during service, the vehicle battery power may have to be interrupted which may "dump" some of this memory. If that happens, then the customer must assume some responsibility for restoring some of these programs. It will prove to be far less of a hassle for you if you get this electronic fact of life understood up front, rather than have to face a surprised and possibly irate customer later.

The tool companies now offer devices designed to be plugged into the cigarette lighter or hooked in parallel to the battery to prevent this in most cases. But, a stated warning, preferably signed by the customer, will prevent wrath of the unwarned customer in the event that some erasure occurs.

Wire Color Codes

Black	BLK
Blue	BLU
Brown	BRN
Dark	DK
Gray	GRY
Green	GRN
Light	LT
Orange	ORN
Pink	PNK
Purple	PPL
Red	RED
Tan	TAN
Violet	VLT
White	WHT
Yellow	YEL

Throughout this manual, the wire color codes listed above are used to help the technician identify which wire is used for what purpose. GM has remained fairly consistent for certain circuits, but not always. In some cases, the technician may have to refer to the specific vehicle diagram to properly identify the wire color.

The common practice used today by GM is to assign a circuit number such as "Circuit 451" or "Circuit 412" to identify a given circuit. This allows the diagnostic "tree" to be applied to different vehicles with different wire colors for the same circuit. The Pin Identifiers and the System Diagrams used in Section IV do not use the circuit numbers. The wire color codes are used to reduce the confusion.

Wire Color Codes

Cont.

The most commonly used wire colors with GM ECM circuits are listed below.

Battery	ORN
Ignition (Switched)	PNK/BLK
Ground	BLK
5V VREF	GRY
Coolant Signal	YEL
MAP Signal	GRN
O^2 Signal	PPL
TPS Signal	BLU
HEI Reference	PNK/WHT
5V Bypass	TAN/BLK
EST	WHT

These color code assignments remain fairly consistent throughout GM applications. Many commonalities occur in the harness wire coloring schemes. Some common sense is required to handle any deviations from the above.

GM Fuel Injection System Evolution

In 1975, GM introduced electronic fuel injection on the new Cadillac Seville. This system was a manifold pressure control system utilizing a Bendix controller built especially for GM. The system had no self-diagnostic subsystem and the temperature sensors were nickel-wire tranducers which are very hard to measure in the field. The throttle body injectors were operated at high pressure. The cold idle up air bypass was a bi-metal device which was not controlled by the ECM. The system was noted for its pep and fuel economy. This system was known as EFI and was used through 1980 on El Dorados and Sevilles. Because the system has little or no correlation to the existing GM designed and manufactured fuel injection systems of today, this manual contains little information on this system.

In 1981, the Cadillac division released a pure GM throttle body fuel-injection system called DEFI. This system had a self-diagnostic sub-system. The system could detect certain circuit faults and the fault codes could be retrieved through the instrument panel display called the On-Board Sentinel. The actual circuit values, however, could not be read through the computer because there was not a complete data stream, as we know it today. Each circuit had to be tapped into directly and tested manually. The DEFI system components are essentially the same as other GM TBI platforms.

In 1982, Cadillac upgraded the DEFI system to a fully digital system called DFI. The ECM used on DFI has a full word data stream, but again, that data is sent to the On-Board Sentinel rather than to the ALDL. The trained technician can read fault codes, sensor values and actuator commands through the Sentinel. The scan tool plugged into the ALDL will not read any of these values. Since the inputs and outputs of both the DEFI and the DFI system are essentially the same as other GM TBI platforms, the Section III circuit troubleshooting information of this manual can be applied to these devices.

Also in 1982, GM introduced the TBI system on engine platforms other than Cadillac. The 2.5L L4 and the 1.8L L4 used a single-point throttle body injection system. The 5.0L and 5.7L V8s used two single-point throttle body injection assemblies in a cross fire configuration. GM divisions used the abbreviations of TBI, EFI or CFI for all of these configurations.

GM Fuel Injection System Evolution

Cont.

In 1984, GM introduced port fuel injection on the 1.8L turbo and the 3.8L V6 engine platforms. This was their first application of airflow controlled multi-port fuel injection. The port fuel injection systems application was expanded the following year to most of the V6 and V8 engines. In 1986, Chevrolet released the tuned-port version which featured specially-tuned intake runners and a Bosch hot wire mass airflow sensor. The truck engines were offered with throttle body style injection beginning in 1988. In 1989, the Chevy-Pontiac-Canada Group deviated from the GM norm by switching from the mass airflow style of EFI to the speed density style of computing fuel requirements. The Buick engines have continued to use mass airflow sensors. However, they are using the Hitachi MAF sensors which seem to be more reliable than the Delco counterparts.

As the year model ascended, the GM ECM became faster, more efficient and more able to maintain engine operation in the event of system failures.

The ECM families are listed later in this section with a more thorough discussion of each type. Since GM vehicles do have a data stream, the technician will want to use the scan tool to access a large amount of information quickly using the scanner connected to the ALDL connector. Just be aware of the fact that the data is only as good as the speed of the data stream. The ability to access the circuits in a "live" test manner is absolutely mandatory. The changes in the software from year to year and occasionally within a model year will sometimes leave holes in the data stream. Also, remember that a misidentified system will result in "No Data" or worse, incorrect data.

To date, GM has used Motorola processors which are very compatible with the scan tools available to the aftermarket. There are several companies which make break-out style interfaces which fit GM ECM connectors. These tools are wonderful for pinpoint circuit testing. GM electronics continues to win worldwide acclaim for design and reliability. It has been said that GM is actually a world-class electronics manufacturer. They just happen to continue making cars so they can sell more electronics.

Operating Modes

The technician must realize that on any electronic fuel injected engine, the computer must be capable of receiving and controlling a variety of inputs and outputs simultaneously throughout a range of engine operating conditions. Therefore, thousands of calculations must be programmed into the engine controller to properly control fuel and timing subsystems as well as other controlled subsystems such as idle and emissions. As the technician uses the individual input and output circuit descriptions in Section III, he will see that the basic operational strategies have been described. Based on the technician's knowledge of engine requirements for starting, idle, acceleration, deceleration, cruise and other operating conditions, he will have to consider the variable factors involved in overall acceptable engine performance. The ECM calculates the amount of fuel, etc. by determining which condition or "mode" in which the engine is operating. The following modes are the possible ECM operating modes.

Starting Mode

For instance, starting an engine requires adjustments for starting conditions. In cold conditions, the computer adjusts the injector pulse width to increase the fuel amount for cold enrichment. As the engine warms up, the pulse width increase delays steadily until normal operating temperature is reached. If an engine is restarted when hot, the computer calculates the pulse width accordingly. The injectors are pulsed once for each distributor reference pulse. The ECM will prime pulse on some engines when the key is turned ON. A few GM engines use a cold start injector system. The cold start injector is electrically controlled through a Thermal Time Switch, not the ECM.

Clear Flood Mode

If the engine becomes flooded, the vehicle operator can instruct the engine control computer to significantly reduce the fuel primer pulses by opening the throttle plate all the way. When the computer "sees" that the TPS signal is at or near "Wide Open Throttle" and the distributor reference frequency is below the run threshold, the computer is programmed to consider that condition as "Clear Flood Request". If the engine is already too flooded to accommodate starting with that fuel reduction, then the technician can disconnect the electrical connection to the injectors and crank the engine with no injector openings at all. Then, once the engine is started, the excess fuel can be removed by revving up the engine and burning away all the excess fuel. The injector harness can be reconnected and the engine started properly, provided the plugs are not too fuel-fouled.

Operating Modes

Cont.

Run Mode – Open Loop

This is the operating mode in which the oxygen sensor is being ignored by the computer. The engine will always be in open loop when starting. If the engine coolant temperature sensor indicates less than normal operating temperature, the computer will ignore the oxygen sensor and run the injector pulse width from its look-up table calculations which ignore Lambda parameters. The system will also operate in open loop whenever the oxygen sensor is not switching above and below the Lambda reference voltage point of approximately 450 millivolts. If the computer "sees" the TPS value at or near the "Wide Open Throttle" value, the computer will again ignore the oxygen sensor and run the injector pulse width according to other sensor input value.

Run Mode – Closed Loop

This is the operating mode in which the computer is correcting the injector pulse width to accommodate the oxygen sensor input. Theoretically, as the oxygen voltage switches above Lambda, indicating rich, the computer adjusts the pulse width down, which should drive the fuel delivered down, causing the oxygen voltage to decrease below the 450 millivolt level. The reduced oxygen voltage value, indicating lean, should then cause the computer to increase the pulse width, allowing more fuel to be delivered, which should result in an increase in oxygen voltage. This toggle of the oxygen sensor voltage and the resultant correction of the injector pulse width should be cyclical, reflecting a normal operating closed loop condition. Remember, speed-density electronic fuel injection systems compute pulse width mostly from engine speed and engine load (MAP) and Air-Flowed controlled engines calculate fuel requirements from MAF.

Acceleration Enrichment Mode

This mode represents the additional programmed pulse width adjustments and timing changes contained in the engine computer's program. These are activated when the computer "sees" a sudden increase in TPS and a sudden increase in MAP pressure. Activation of this mode should immediately result in a drastic increase in injector pulse width. This maximum increase will vary from one type of fuel system to the next, but will typically be in the area of a ten millisecond pulse width or greater. The TBI engines and some PFI engines can also switch from synchronous to asynchronous pulses if the acceleration demand is great.

Operating Modes

Cont.

Deceleration Leanment Mode

When a vehicle is decelerating, the only fuel actually required is the amount of fuel necessary to prevent the engine from stalling. Since decelerating the engine increases the manifold vacuum, any fuel input will be drawn into the combustion chambers and wasted, producing essentially no power. Therefore, this mode allows the computer to decrease the injector pulse width to a point where only the essential quantity of fuel is administered to prevent the engine from stalling. Some ECMs will actually enter another mode called Decel Fuel Cut-Off if deceleration is drastic enough.

Default Values Mode

If the computer receives no signal or an out-of-range signal from any of its key input sensors, the computer is programmed to put the system into what GM calls the "Default Values" Mode. The computer will substitute one or more of the other sensor values to facilitate the best engine operation it can. On the early systems, Default was obvious. The Check Engine lamp was ON, and the vehicle ran very poorly. On the later systems, the Check Engine lamp will be ON, but the backup programs have been so improved, the vehicle will quite often run very well. Remember, the Check Engine lamp can be illuminated when certain emission systems have malfunctioned and this does not necessarily mean that the system is in the Default Values Mode. In some cases the vehicle will run better than normal and the performance will be improved. This will be short term because the engine will not run in this enriched condition indefinitely. The fuel economy will also suffer. It is interesting to note that on later systems, virtually all of the key input signals can be lost and the default system values will usually keep the vehicle operating. The one exception is the Reference signal. No Reference–No Run!

Backup Mode

In this mode, the ECM has detected a major control malfunction which will not allow normal pulse control. The ECM is running on the Calpac, which is a chip specifically designed for "limp-in" in case of a major ECM malfunction.

In summary, the technician must be able to recognize the engine's operating mode in order to correctly diagnose problems with GM fuel injected systems. While stall testing or road testing a GM vehicle, learn to recognize and verify proper system control in these operating modes.

GM System ID

Typical GM Serial Number

1	G	3	A	P	8	7	2	4	C	L	1	7	3	1	7	3
(1)	(2)	(3)	(4)	(5)	(6)	(7)	(8)	(9)	(10)	(11)	(12)	(13)	(14)	(15)	(16)	(17)

Position 1: Nation of Origin
- 1 - USA
- 2 - Canada
- J - Japan

Position 2: Manufacturer
- G - General Motors
- S - Suzuki
- 8 - Isuzu
- Y - Nummi

Position 3: Division
- 1 - Chevrolet
- 2 - Pontiac
- 3 - Oldsmobile
- 4 - Buick
- 5 - Cadillac
- 7 - GM of Canada
- 8 - Truck
- T - Truck

Position 4 and 5: Car line and Series (Not important to engine testing.)

Position 6: Body Style (Again not relevant to engine testing.)

Position 7: Restraint Code (What type of seat belts?) (Not relevant)

Position 8: Engine Code (Very important, varies with model year.)

Position 9: Check Digit (Not relevant.)

GM System ID (Cont.)

Position 10: Model Year

B - 81	G - 86
C - 82	H - 87
D - 83	J - 88
E - 84	K - 89
F - 85	L - 90

Position 11: Plant Code (Not used for engine testing.)

Obviously, the most important digits are the 8th, engine size, and the 10th, model year. The technician should use these for every diagnostic report. These two identifications are absolutely necessary in order to utilize the scanner tool, select the proper diagnostic trees and verify the application of wiring or service bulletin information.

Example: Using the serial number listed on the previous page, the technician can tell that this is a GM car, built in the US, it is an Olds, 1982 with a "2" engine which the 82 engine charts will tell us that a "2" engine is 2.5L L4. Using the system ID chart, the technician can then tell that the computer system used would be EFI-TBI. This will get the technician thinking about a standard TBI System, low pressure 9-13 PSI, with speed-density system or manifold pressure control for main fuel computations.

The following pages provide the major identification information for GM EFI engines from 1981 through 1990. The abbreviations used in the ID Charts and their definitions are listed below.

T - Turbo

HEI - High Energy Ignition

DIS - Direct Ignition System

IDI - Integrated Direct Ignition

CCCI - Computer Controlled Coil Ignition

The Computer abbreviations are described following the ID charts.

GM System ID

Year (10th)	VIN (8th)	Engine	Computer Type	Ignition
1981	6	6.0L V8	DEFI	HEI
1982	0	1.8L L4	EFI-TBI	HEI
	2	2.5L L4	EFI-TBI	HEI
	7	5.0L V8	CFI-TBI	HEI
	8	5.7L V8	CFI-TBI	HEI
	8	4.1L V8	DFI	HEI
	9	6.0L V8	DFI	HEI
	R	2.5L L4	EFI-TBI	HEI
1983	0	1.8L L4	EFI-TBI	HEI
	2	2.5L L4	EFI-TBI	HEI
	8	5.7L V8	CFI-TBI	HEI
	8	4.1L V8	DFI	HEI
	9	6.0L V9	DFI	HEI
	P	2.0L L4	EFI-TBI	HEI
	R	2.5L L4	EFI-TBI	HEI
	S	5.0L V8	CFI-TBI	HEI
1984	0	1.8L L4	PONT. L4 EFI	HEI
	2	2.5L L4	PONT. L4 EFI	HEI
	3	3.8L V6	BUICK 6 PFI	HEI
	8	5.7L V8	CFI-TBI	HEI
	8	4.1L V8	DFI	HEI
	9	6.0L V8	DFI	HEI
	9	3.8L V6-T	BUICK 6 PFI	CCCI
	J	1.8L L4-T	PONT.4 PFI	HEI
	P	2.0L L4	EFI-TBI	HEI
	R	2.5L L4	PONT.4 EFI	HEI
1985	0	1.8L L4	PONT. L4 EFI	HEI
	2	2.5L L4	PONT. L4 EFI	HEI
	3	3.8L V8	BUICK 6 PFI	HEI
	8	5.7L V8	CHEV. PFI	HEI
	8	4.1L V8	DFI	HEI
	9	2.8L V6	PONT. 6 PFI	HEI
	9	3.8L V6-T	BUICK 6 PFI	CCCI
	F	5.0L V8	CHEV. PFI	HEI
	E	2.5L L4	PONT. L4 EFI	HEI
	J	1.8L L4-T	PONT. 4 PFI	HEI
	L	3.0L V6	BUICK 6 PFI	CCCI
	P	2.0L LR	EFI-TBI	HEI
	R	2.5L L4	PONT. L4 EFI	HEI
	S	2.8L V6	CHEV.PFI	HEI
	U	2.5L L4	PONT. L4 EFI	HEI

GM System ID

Year (10th)	VIN (8th)	Engine	Computer Type	Ignition
1985 cont.	W	2.8L V6	CHEV.PFI	HEI
	Z	4.3L V6	CFI 2BL TBI	HEI
	U	2.5L L4	PONT. L4 EFI	HEI
1986	0	1.8L L4	PONT. L4 EFI	HEI
	1	2.0L L4	EFI-TBI	HEI
	2	2.5L L4	PONT. L4 EFI	HEI
	3	3.8L V6	BUICK PFI	CCCI
	7	3.8L V6	BUICK PFI	HEI
	8	4.1L V8	DFI	HEI
	8	5.7L V8	TPI V8	HEI
	8	5.7L V8	P4 TPI V8	HEI
	9	2.8L V6	PONT. 6 PFI	HEI
	9	3.8L V6-T	BUICK PFI	CCCI
	B	3.8L V6	BUICK PFI	HEI
	E	2.5L L4	PONT. L4 EFI	HEI
	F	5.0L V8	P4 TPI V8	HEI
	F	5.0L V8	TPI V8	HEI
	J	1.8L L4-T	PONT. 4 PFI	HEI
	L	3.0L V6	BUICK PFI	CCCI
	P	2.0L L4	EFI-TBI	HEI
	R	2.5L L4	PONT. L4 EFI	HEI
	R	2.8L V6	CFI 2BL TBI	HEI
	S	2.8L V6	CHEV. PFI	HEI
	U	2.5L L4	PONT. L4 EFI	HEI
	W	2.8L V6	CHEV. PFI	HEI
	Z	4.3L V6	CFI 2BL TBI	HEI
1987	1	2.0L L4	P4 CHEV.	DIS
	3	3.8L V6	BUICK PFI	CCCI
	7	3.8L V6-T	BUICK PFI	CCCI
	8	4.1L V8	DFI	HEI
	8	5.7L V8	P4 TPI V8	HEI
	8	5.7L V8	TPI V8	HEI
	9	2.8L V6	PONT. 6 PFI	HEI
	E	2.5L L4	P4 TRUCK	HEI
	F	5.0L V8	P4 TPI V8	HEI
	H	5.0L V8	TRUCK TBI	HEI
	K	2.0L L4	P4 PONT.	HEI
	K	5.7L V8	TRUCK TBI	HEI
	L	3.0L V6	BUICK PFI	CCCI
	M	2.0L L4	P4 TURBO	HEI
	N	7.4L V8	TRUCK TBI	HEI
	R	2.5L L4	P4 PONT.	DIS
	R	2.8L V6	CFI 2BL TBI	HEI

GM System ID

Year (10th)	VIN (8th)	Engine	Computer Type	Ignition
1987 cont.	S	2.8L V6	CHEV. PFI	HEI
	U	2.5L L4	P4 PONT.	DIS
	W	2.8L V6	P4 PFI V6	DIS
	Z	4.3L V6	CFI 2BL TBI	HEI
1988	1	2.0L L4	P4 CONT. 4	DIS
	3	3.8L V6	BUICK PFI	CCCI
	5	4.5L V8	4500	HEI
	6	1.6L L4	P4 TBI V8	HEI
	8	4.5L V8	4500	HEI
	8	5.7L V8	P4 TPI V8	HEI
	8	5.7L V8	TPI V8	HEI
	9	2.8L V6	PONT. 6 PFI	HEI
	C	3.8L V6	P4 3800	CCCI
	D	2.3L L4	P4 QUAD 4	IDI
	E	5.0L V8	CFI 2BL TBI	HEI
	E	2.5L L4	P4 TBI L4 TRUCK	HEI
	F	5.0L V8	P4 TPI V8	HEI
	K	2.0L L4	P4 CONT. 4	HEI
	L	3.0L V6	BUICK PFI	CCCI
	M	2.0L L4-T	P4 TURBO	HEI
	N	7.4L V8	TRUCK TBI	HEI
	R	2.5L L4	P4 CONT. 4	DIS
	R	2.8L V6	CFI 2BL TRUCK	HEI
	S	2.8L V6	CHEV. PFI	HEI
	T	3.1L V6	P4 V6 MAP	HEI
	T	3.1L V6	P4 V6 MAP	HEI
	W	2.8L V6	P4 PFI V6	DIS
	W	2.8L V6	P4 V6 MAP	DIS
	U	2.5L L4	P4 CONT. 4	DIS
	Z	4.3L V6	CFI 2BL TBI	HEI
1989	1	2.0L L4	P4 CONT.4	DIS
	5	4.5L V8	4500	HEI
	6	1.6L L4	P4 TBI L4	HEI
	7	3.8L V6	BUICK PFI	CCCI
	7	5.7L V8	CFI 2BL TBI	HEI
	8	5.7L V8	P4 TPI V8	HEI
	A	2.3L L4	QUAD 4	IDI
	C	3.8L V6	P4 3800	CCCI
	D	2.3L L4	QUAD 4	IDI
	E	5.0L V8	CFI 2BL TBI	HEI
	E	2.5L L4	P4 TBI TRUCK	HEI
	F	5.0L V8	P4 TBI V8	HEI
	H	5.0L V8	TRUCK TBI	HEI

GM System ID

Year (10th)	VIN (8th)	Engine	Computer Type	Ignition
1989 cont.	J	5.7 V8	P4 LOTUS	HEI
	K	2.0L L4	P4 CONT. 4	DIS
	K	5.7L V8	TRUCK TBI	HEI
	M	2.0L L4-T	P4 TURBO	HEI
	N	7.4L V8	TRUCK TBI	HEI
	N	3.3L V6	P4 3300	CCCI
	R	2.5L L4	P4 CONT.4	DIS
	R	2.8L V6	CFI 2BL TBI TRUCK	HEI
	S	2.8L V6	CHEVY PFI	HEI
	T	3.1L V6	P4 V6 MAP	DIS
	U	2.5L V6	P4 CONT. 4	DIS
	W	2.8L V6	P4 V6 MAP	DIS
	Z	4.3 V6	CFI 2BL TBI	HEI
	Z	4.3 V6	TRUCK TBI	HEI
1990	3	4.5L V8	4500	HEI
	6	1.6L L4	P4 TBI L4	HEI
	7	5.7L V8	CFI 2BL TBI	HEI
	8	5.7L V8	P4 TBI V8	HEI
	8	4.5L V8	4500	HEI
	A	2.3L L4	P4 QUAD 4	IDI
	C	3.8L V6	P4 3800	CCCI
	C	3.8L V6	GM 30	CCCI
	D	2.3L L4	P4 QUAD 4	IDI
	E	5.0LV8	CFI 2BL TBI	HEI
	E	2.5L L4	P4 TBI L4	HEI
	F	5.0L V8	P4 TPI V8	HEI
	G	2.2L L4	PONT. 4PFI	DIS
	H	5.0L V8	TRUCK TBI	HEI
	K	2.0L L4	P4 TBI L4	DIS
	K	2.0L L4	P4 CONT. 4	DIS
	K	5.7L V8	TRUCK TBI	HEI
	M	2.0L L4-T	P4 TURBO	HEI
	M	7.0L V8	TRUCK TBI	HEI
	N	3.3L V6	P4 3300	CCCI
	N	7.4L V8	TRUCK TBI	HEI
	R	2.5L L4	P4 CONT.4	DIS
	R	2.5L L4	GM 10	DIS
	R	2.8L V6	CFI 2BL TBI	HEI
	T	3.1L V6	P4 V6 MAP	DIS
	U	2.5L L4	P4 CONT.4	DIS
	V	3.1L V6-T	31 TURBO	DIS
	Z	4.3L V6	TRUCK TBI	HEI
	Z	4.3L V6	CFI 2BL TBI	HEI

GM ECM Controller Types

C-3 System Descriptions - Slow Speed - 160 Baud Rate

EFI-TBI: This ECM family is used on vehicles with throttle body injection. Introduced in late 1982, it is still used to date on many cars, although most cars since late 1987 now have the newer P4 systems.

CFI 2B TBI: This ECM family was originally used on Corvettes, Camaros and Trans Ams with two single injector throttle body fuel injection assemblies. It is also used on throttle body systems with one twin injector throttle body assembly.

PONT 4 PFI: This ECM family is used on many of the four cylinder engines with port fuel injection or one injector at each port. Some car divisions named this type of system MFI (Multi-Port Fuel Injection).

BUICK 6 PFI: This ECM family also controls injectors at each intake port, yet it is designed for six-cylinder applications. It can also be called MFI, and when used to fire each injector individually in sequence, it is called SFI (Sequential Fuel Injection).

PONT L4 EFI: This ECM family is used on Pontiac engines equipped with throttle body injection.

CHEV PFI: This ECM family is used on certain Chevrolet engines (not necessarily Chevrolet cars) which have injectors at each port. It can also be called MFI (Multi-Port Fuel Injection).

PONT 6 PFI: This ECM family is used on certain six-cylinder Pontiac engines that have port fuel injection.

BUICK PFI: This ECM family was another name for the Buick 6 PFI. The difference is primarily revision levels of the software utilized to initiate the system inputs and controls.

GM ECM Controller Types

Cont.

TPI V8: This ECM family is used on the 5.0 and 5.7L Chevrolet engines which have the special air intake plenum which has been "tuned" or made with special flow characteristics. It is a port fuel type system known for its high performance. Its normal serial data baud rate is 160, yet with special adaptors, some scanners can read another serial line at the high speed baud rate of 8192.

TRUCK TBI: This family of ECMs is used on Chevrolet and GMC trucks which have throttle body fuel injection.

T SPECTRUM: This ECM family is very special. It is used on the Chevrolet 1.5L L4 turbocharged multi-port fuel injection Spectrum car. This application required a special adaptor cable to read the data stream.

P-4 System Descriptions - High Speed - 8192 Baud Rate

P-4 PONT: This ECM family is used on certain late model Pontiac four-cylinder engines with throttle body injection.

P-4 TURBO: This ECM family is used on the 2.0L turbocharged Pontiac engine with port fuel injection. The same TPI adaptor cable used on Chevrolet TPI may be required to read the high speed data transmission.

P-4 CHEV: This ECM family is used on the Chevrolet 2.0L engine with throttle body injection.

P-4 PFI V6: This ECM family is used on the Chevrolet 2.8L V6 late model engines. The special adaptor cable may also be required to read this serial stream.

P-4 TPI V8: This ECM family is for the late model Tuned Port Injection systems used on the 5.0L and 5.7L engines. The TPI adaptor cable is required to read this data line.

P-4 TRUCK: This ECM family is used when the 2.5L Pontiac engine is used in a truck application.

GM ECM Controller Types

Cont.

P4T - SPECTRUM: This ECM family is used in the 1987 1.2L and up Spectrum Turbo applications. You must use the special adaptor cable in order to access the data stream.

P-4 TBI L4: This ECM family is new for 1988 through 1990 models with the 2.5L L4 in the S trucks and the 1.8 L4 in the Pontiac Lemans.

P-4 CONT 4: This ECM family is the new high speed unit used from 1988 through 1990 on both Chevrolet and Pontiac four cylinder throttle body fuel injection applications.

QUAD 4: This ECM is dedicated to controlling the inputs and outputs for the new Olds four cylinder OHC multi-port fuel injected engine known as the Quad 4.

P-4 V6 MAP: This ECM family is used for the 1988 through 1990 Chevrolet 2.8L V6 engines that used a MAP sensor along with a MAF sensor.

P-4 3800: This ECM family is for the new 1988 and up 3.8L Buick engine known as the 3800. This engine is ported and fires each injector sequentially so it is called SFI.

P-4 3300: This ECM is for the updated 3.3L V6 Buick engine.

4500: This ECM family is used on the 1988-1990 4.5L V8 Cadillacs. It still has the name DFI. This system also has a body control module that is controlled by the ECM (Engine Computer Module).

CFI 2BL TBI: This ECM family is used for 1988-1990 Chevy and Pontiac engines. It features two single-barrel throttle bodies mounted on a crossfire plenum, or it may have a two-barrel throttle body with two injectors mounted on it. This ECM family is the old CFI combined with the newer style two-barrel throttle body.

P-4 Lotus: This ECM family is used as an optional system on 1989 Corvettes with a 5.7L port fuel injection. This system has a Lotus/Chevrolet engine.

TBI Engines

The 1.8L L4 and 2.0L L4 Chevrolet engines are used by various care divisions as a throttle body injected systems. Later models of the 1.8L engine are available as turbocharged port fuel engines. They are called EFI-TBI, P-5 CHEV or P-4 CONT 4. The early models have GM's Model 300 or 500 Throttle Body. The later models have the GM Model 700 Throttle Body.

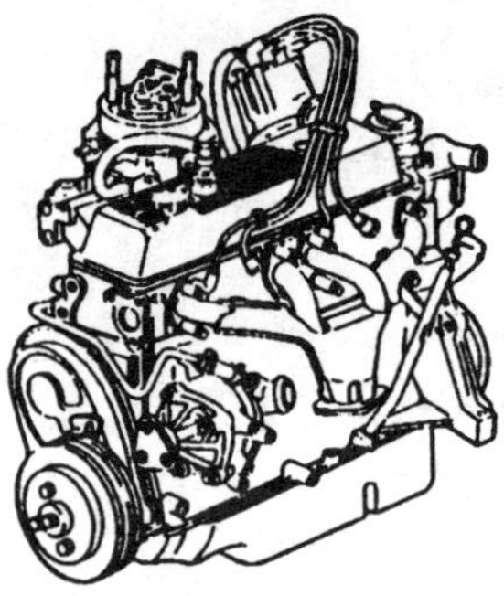

The 2.5L L4 Pontiac engine is used by various car and truck models as a throttle body injected system. It is also called EFI-TBI of PONT L4 TBI, P4 PONT, P-4 TRUCK or P-4 CONT 4. The early models used the GM Model 300 Throttle Body Assembly. The later models use the GM Model 700 TBI Assembly.

The 4.3L V6, 5.0L and the 5.7L V8s have been offered with both throttle body injection and with port fuel injection. The original conversion to fuel injection was a configuration for the V8s called Cross-Fire injection. This system used the GM Model 400 Throttle Body Assemblies. Each assembly contained one injector which delivered fuel to four of the engine's cylinders through the specially-designed manifold. Later configurations for the 5.0L, 5.7L and 4.3L V6 use the twin-bore throttle body assembly containing two injectors each firing alternately and subsequently fueling opposing groups of cylinders.

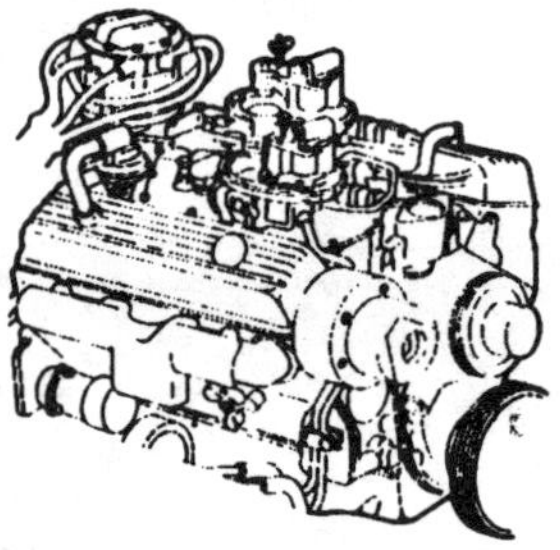

PFI Engines

The 1.8L L4 engine pictured in Figure 4 was originally introduced as a throttle body fueled engine. In 1986, it became a turbocharged port fuel engine. In 1987, it was bored and stroked to 2.0L and is offered as a turbocharged port fueled engine. It is called PONT 4 PFI or P4 TURBO.

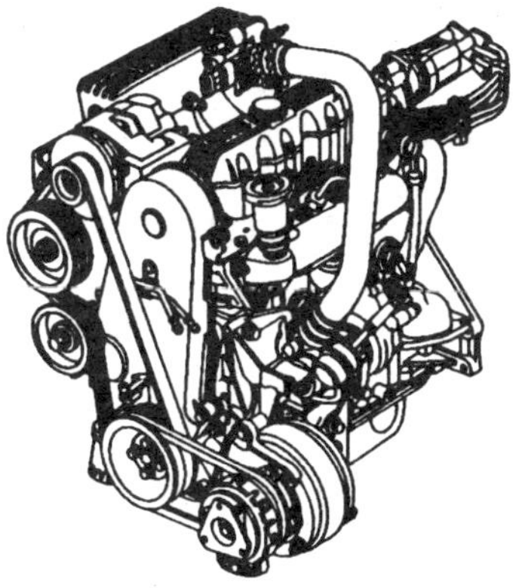

Figure 4

Figure 5

The 2.8L V6 engine illustrated in Figure 5 is equipped with two types of port fuel injection. It can be equipped with a MAP sensor and that system is a Speed-Density type of fuel control system. That system is called CHEV PFI. The 2.8L generation II engine has a different intake manifold and fuel rail. In 1989, the new 3.1L version was released. It is generally a Speed-Density type.

The 3.0L is always configured with port fuel injection. Figure 6 illustrates that type of engine. This system fires the injectors as a group once for each crankshaft revolution. It uses a MAF sensor. It is called BUICK 6 PFI or BUICK PFI. In 1989, the 3.3L or "3300" version was released. It is still a MAF controlled simultaneous double-fire system.

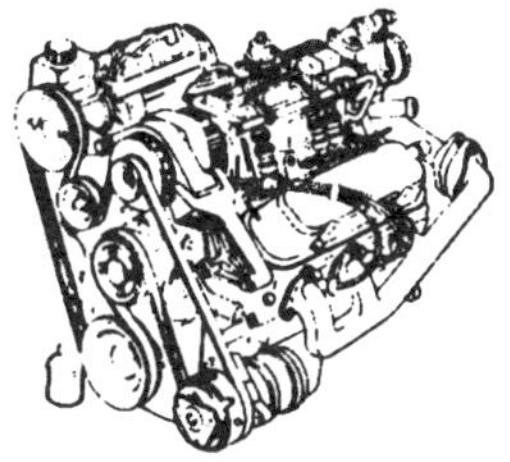

Figure 6

PFI Engines (Cont.)

The 3.8L V6 engine pictured in Figure 7 is naturally aspirated and is only offered with port fuel injection. It also features simultaneous double-fire injection. It is identified as BUICK 6 PFI or BUICK PFI.

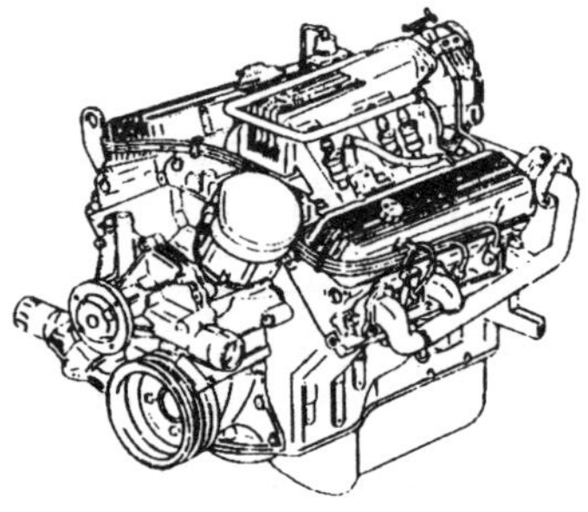

Figure 7

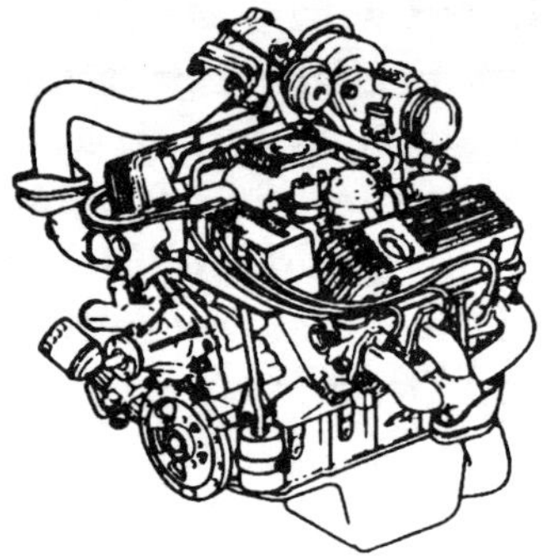

Figure 8

Figure 8 illustrates the same 3.8L V6 as it appears with turbo charging. In this application it can be fired sequentially or one injector after another. It is called BUICK 6 PFI or BUICK PFI. For 1988, it becomes the new 3800 V6 and its system is called PR3800.

The 5.0/5.7L V8 engines are configured with throttle body-type fuel injection for most applications but are also offered with port fuel calle Tuned Port Injection. This system features precise air intake runners to maximize volumetric efficiency without turbo charging. This system is called CHEV PFI or TPI V8 or P4 TPI V8.

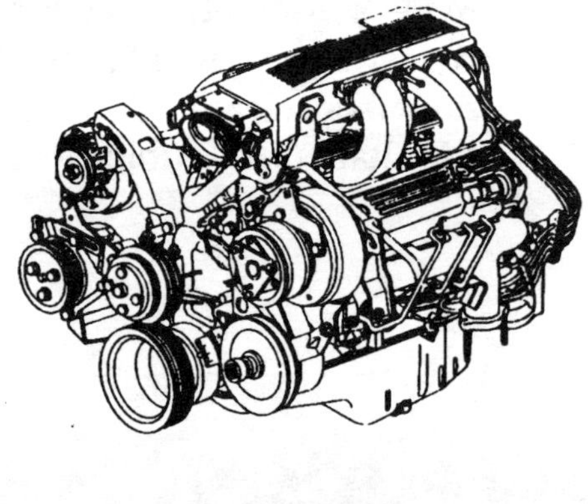

Figure 9

Fuel Rail Configuration

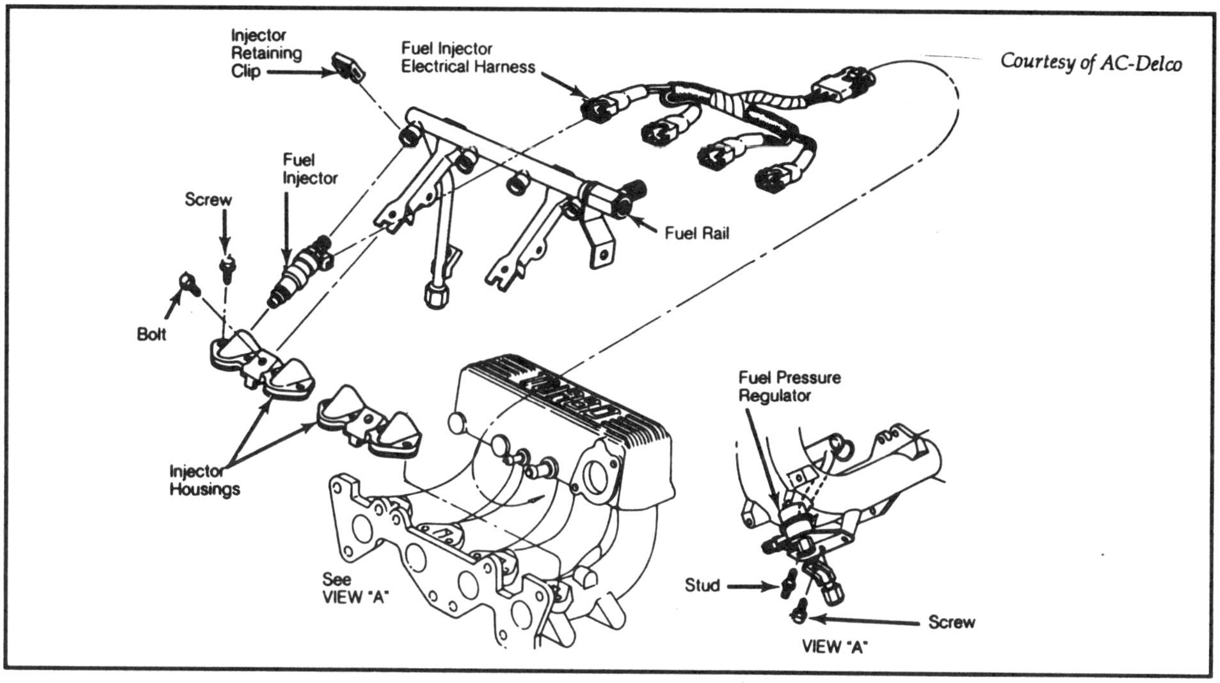

Typical 1.8L L4 Fuel Rail

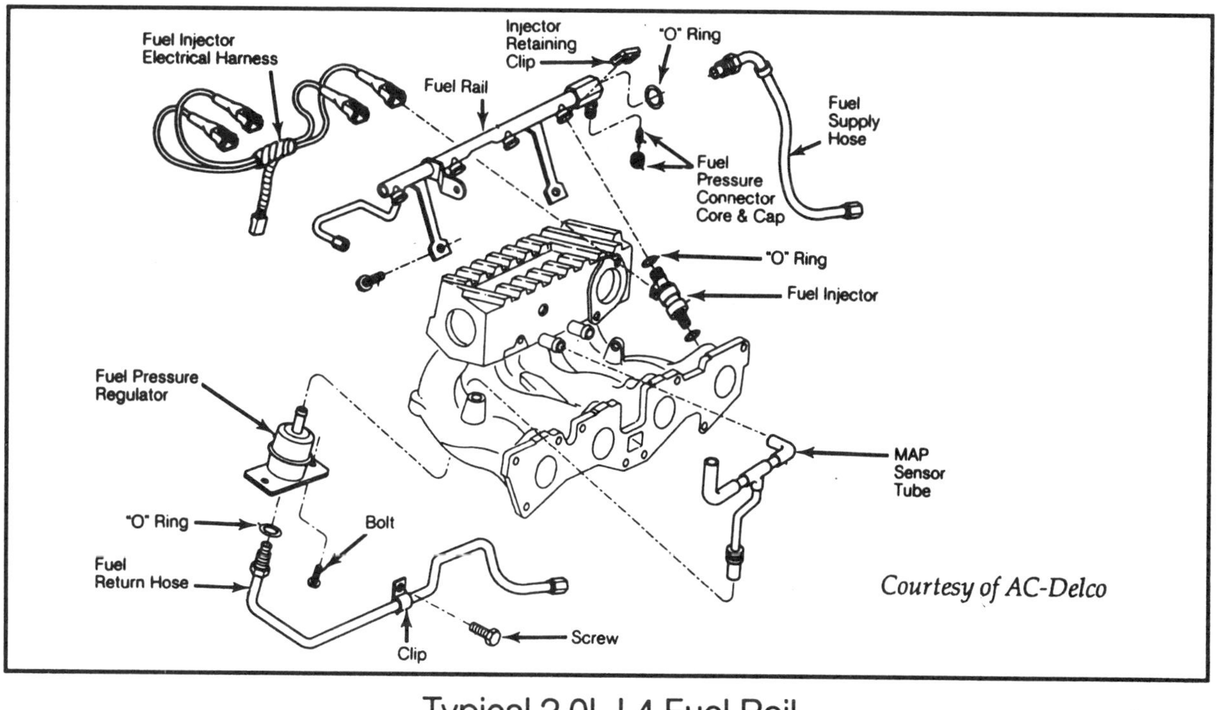

Typical 2.0L L4 Fuel Rail

Fuel Rail Configuration (Cont.)

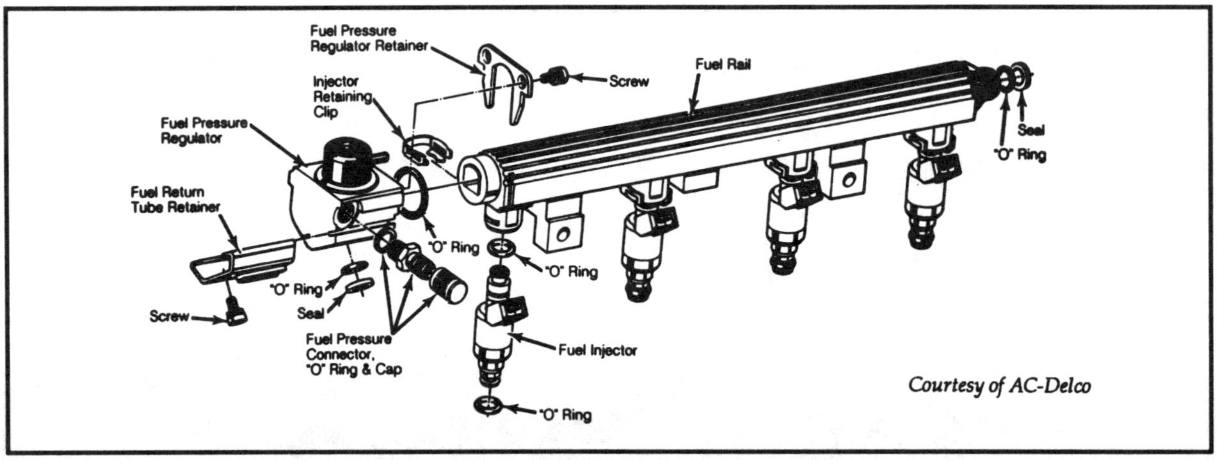

Typical 2.3L Quad 4

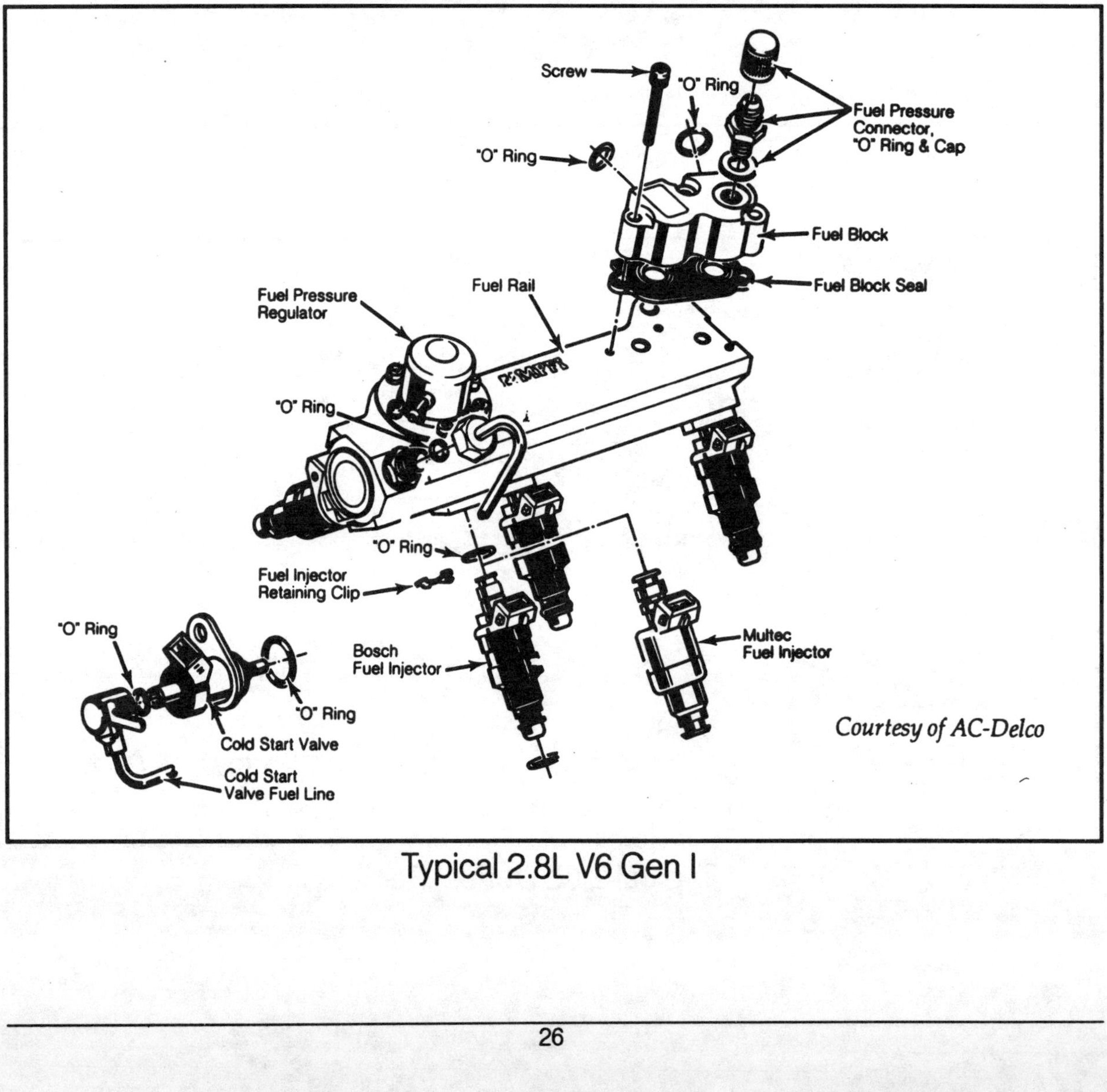

Typical 2.8L V6 Gen I

Fuel Rail Configuration (Cont.)

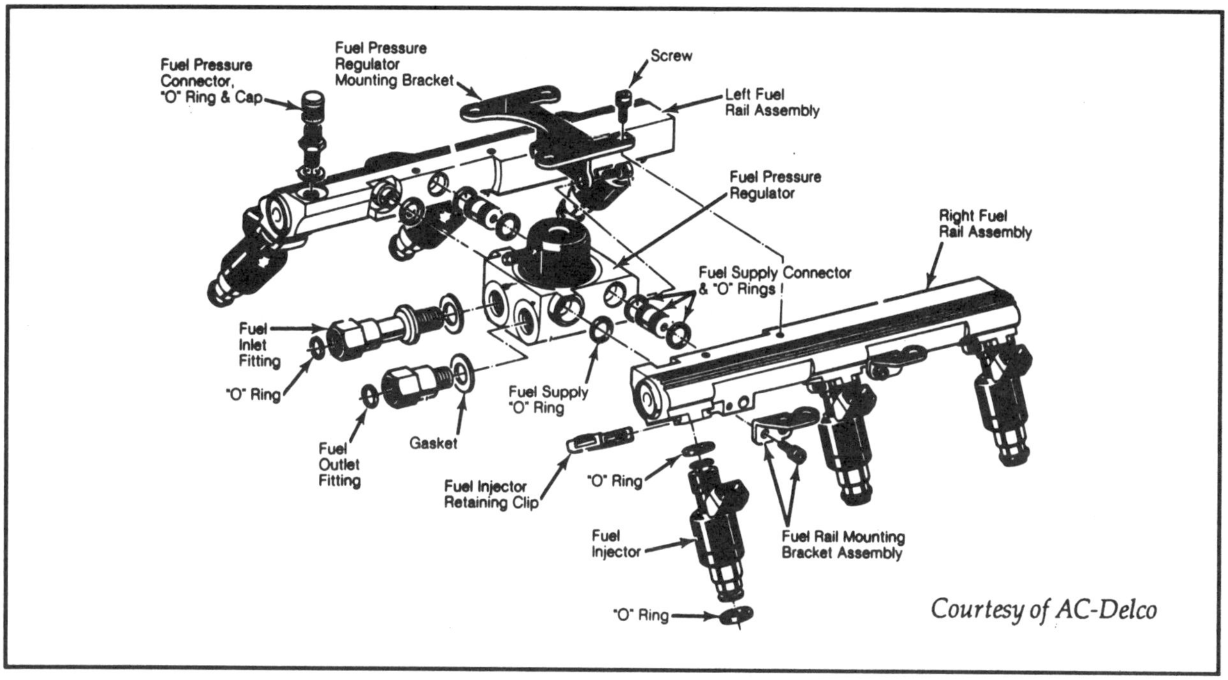

Typical 2.8L V6 Gen II

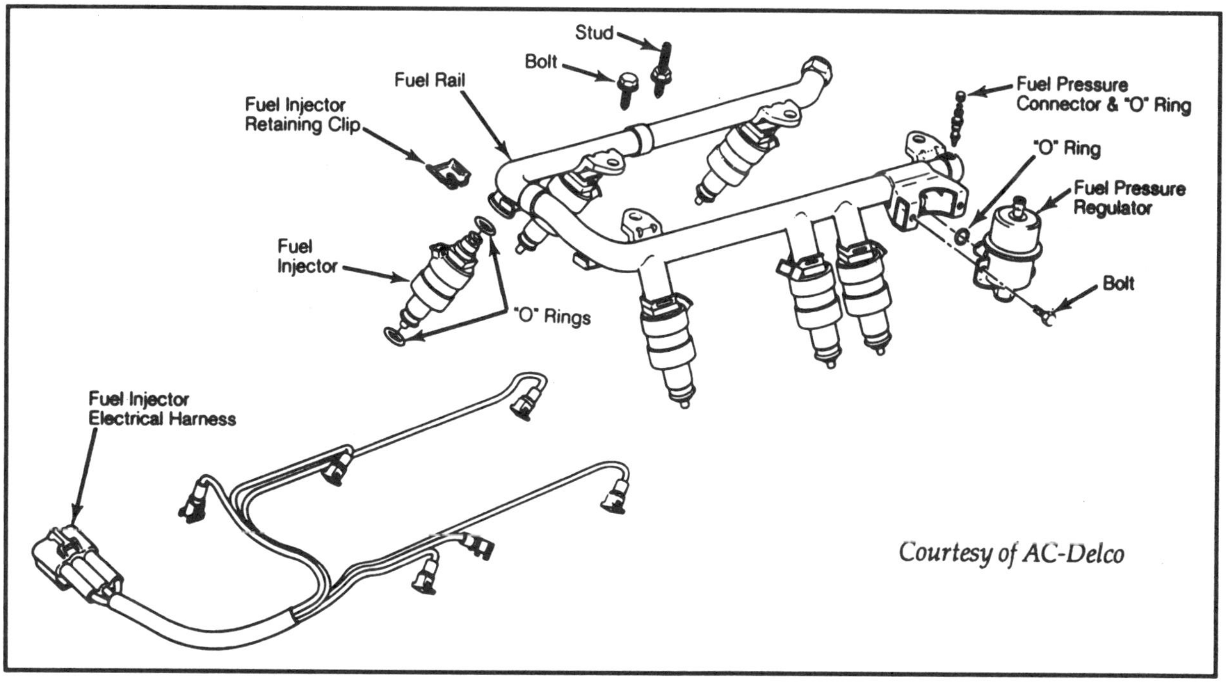

Typical 3.0L V6

Fuel Rail Configuration (Cont.)

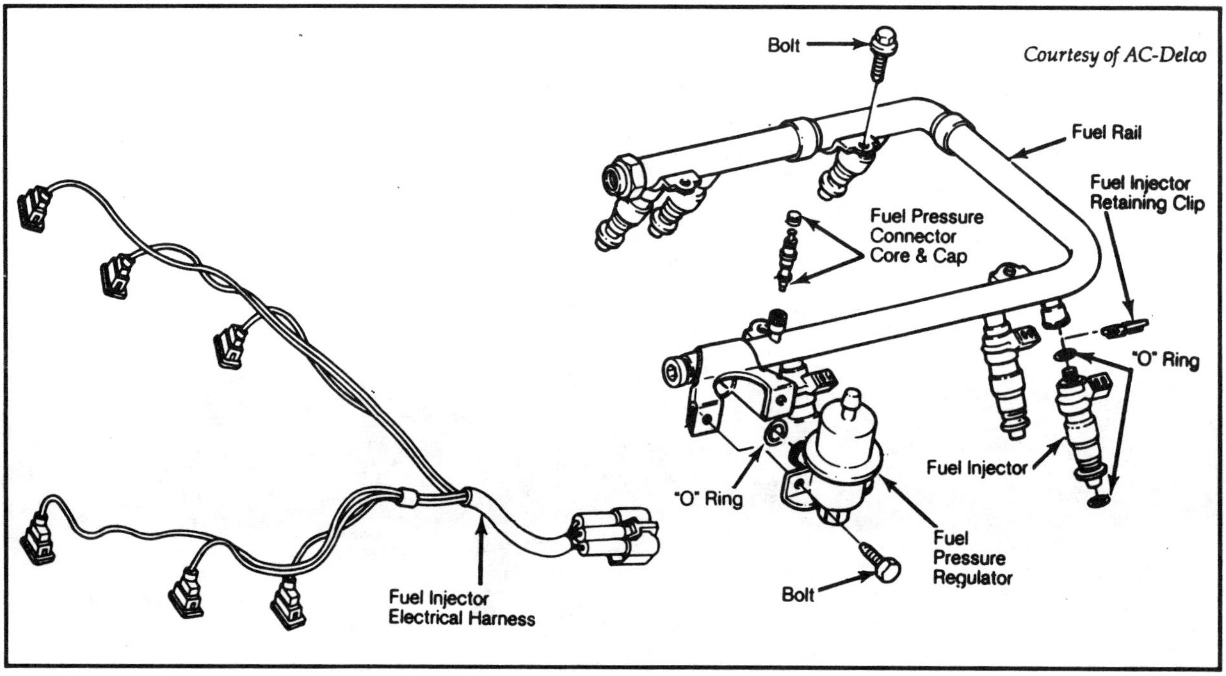

Typical 3.8L V6

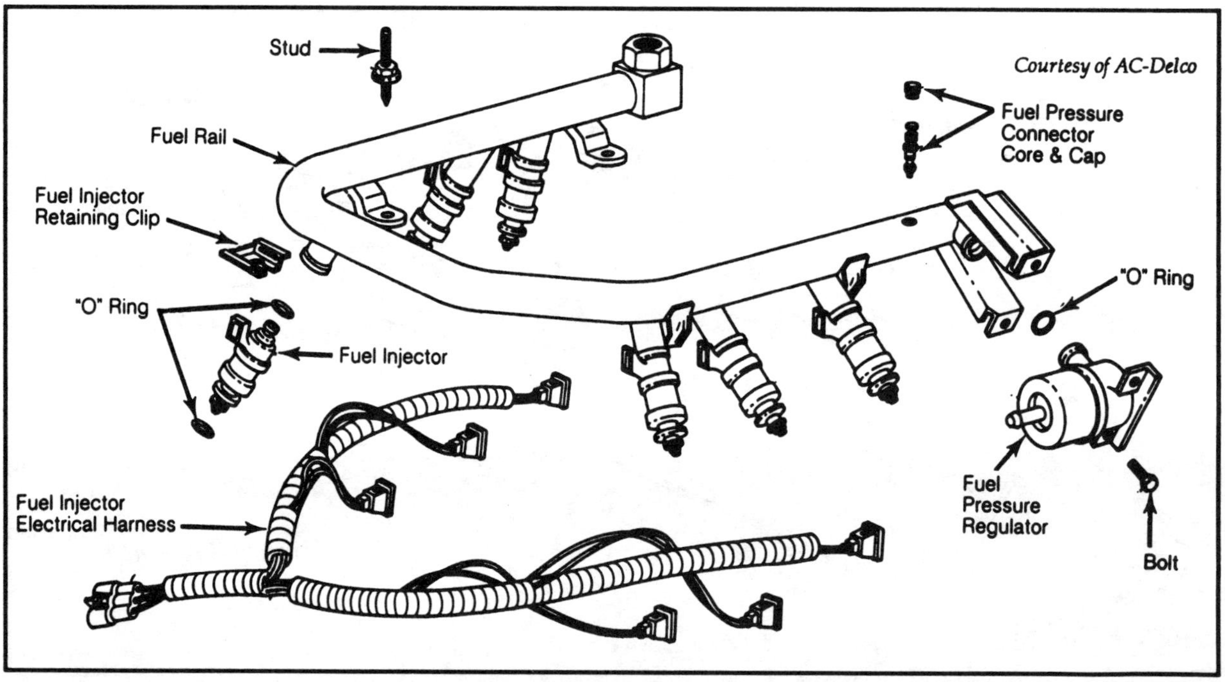

Typical "3800" V6

Fuel Rail Configuration (Cont.)

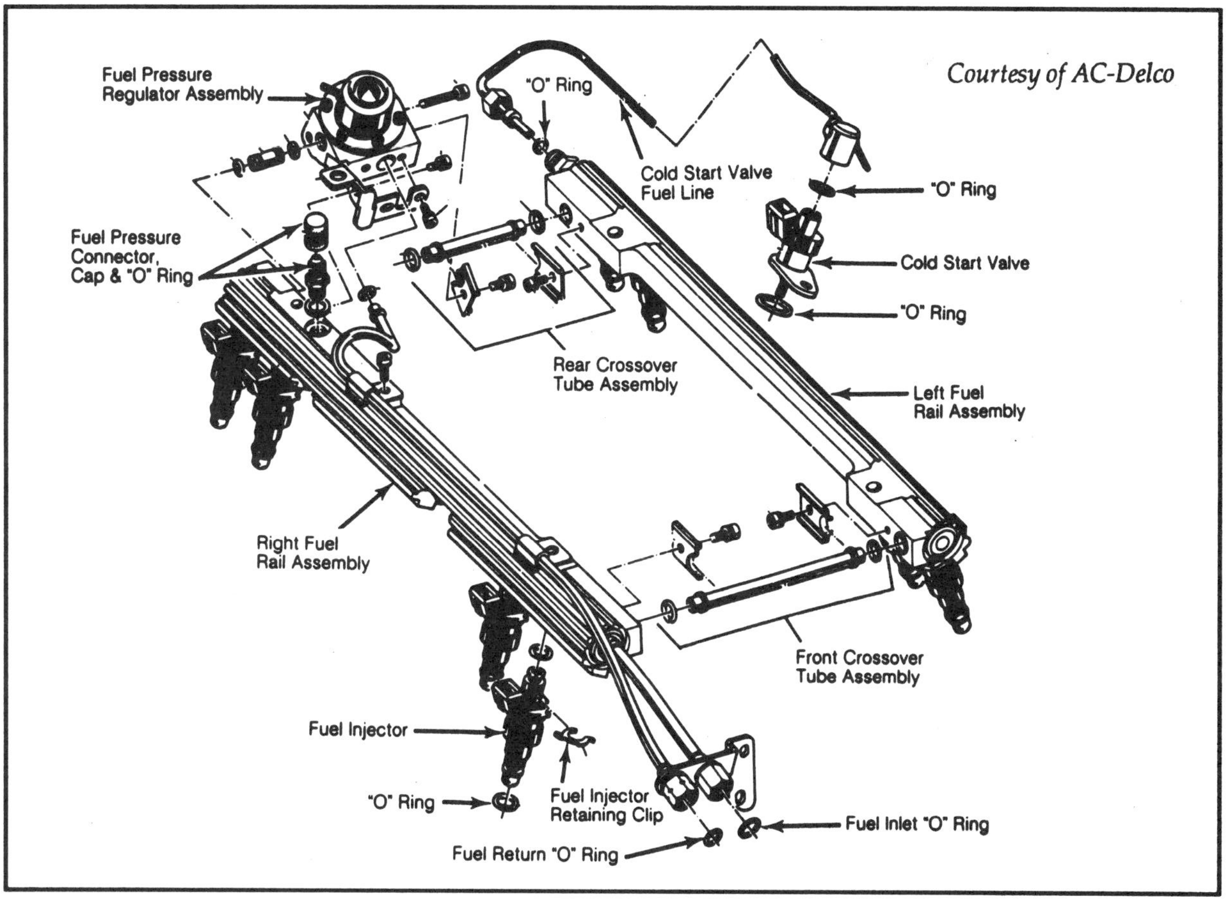

Typical 5.0L & 5.7L V8

PFI Throttle Bodies

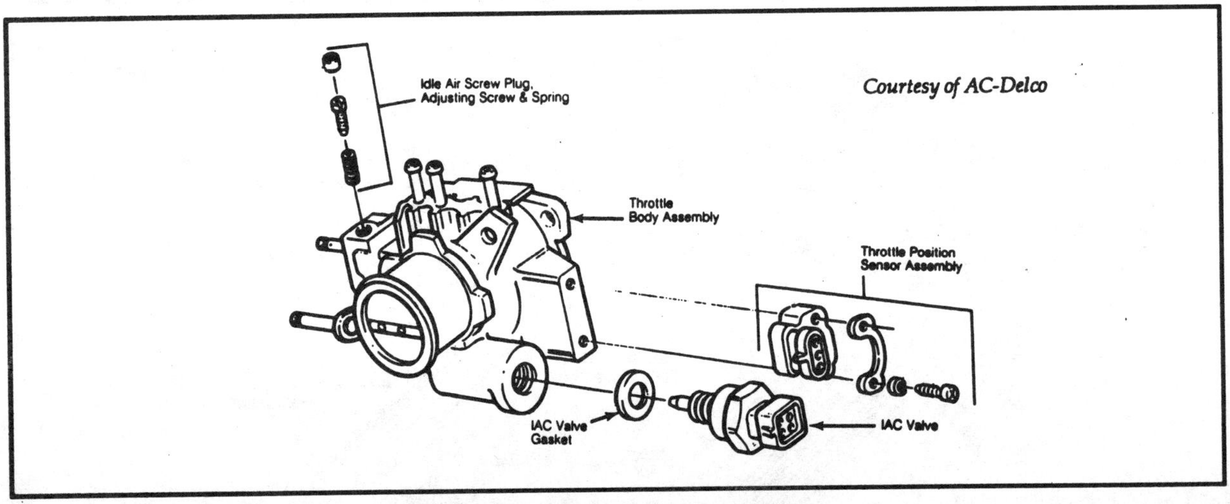

Typical 1.8L Throttle Body

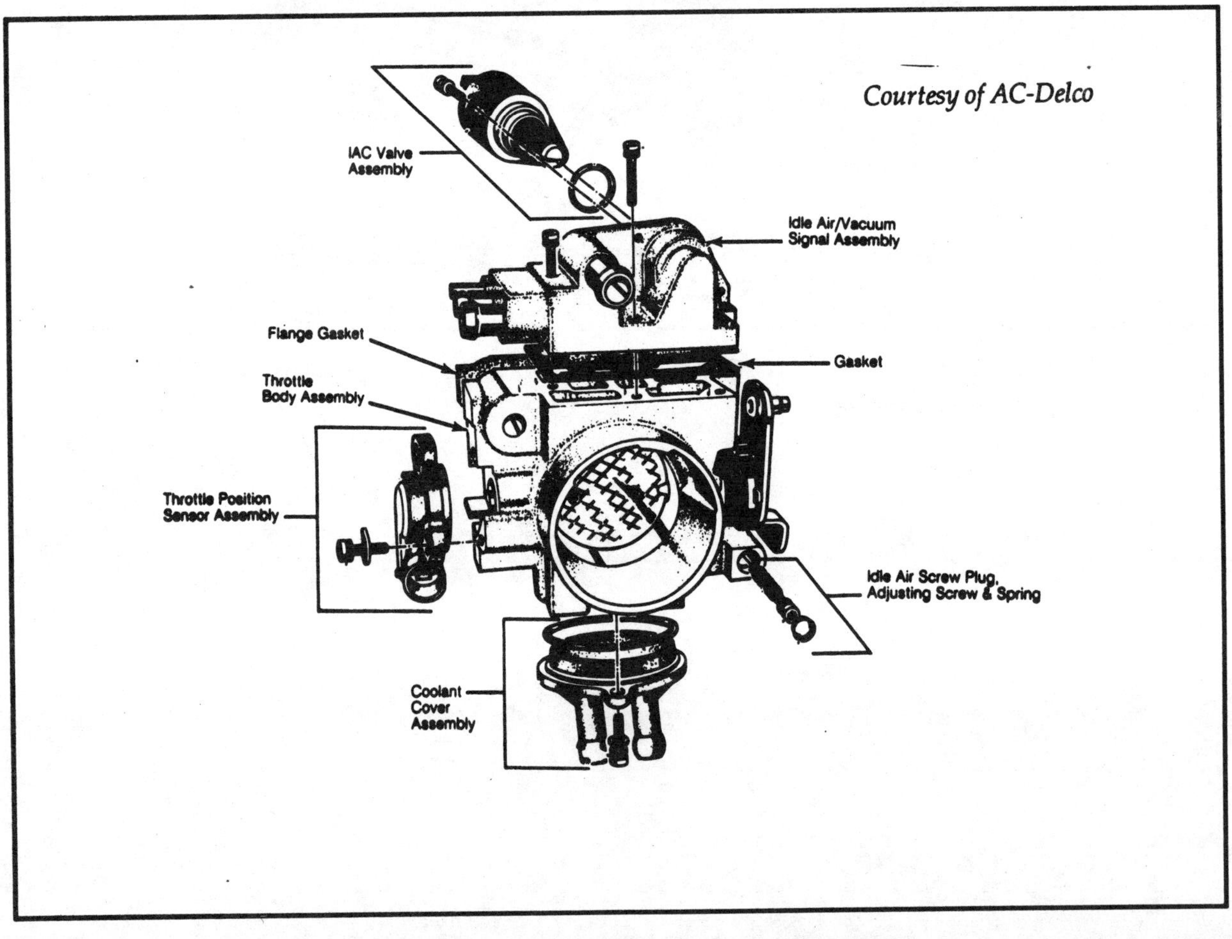

Typical 2.3L "QUAD 4" Throttle Body

PFI Throttle Bodies (Cont.)

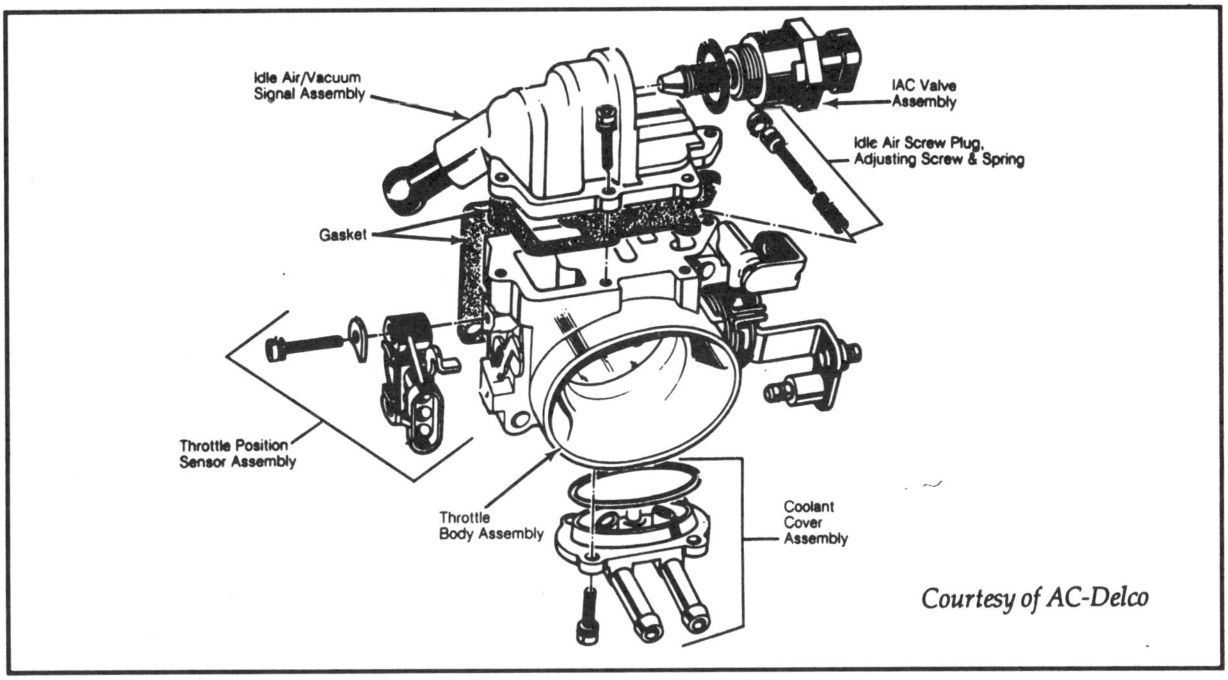

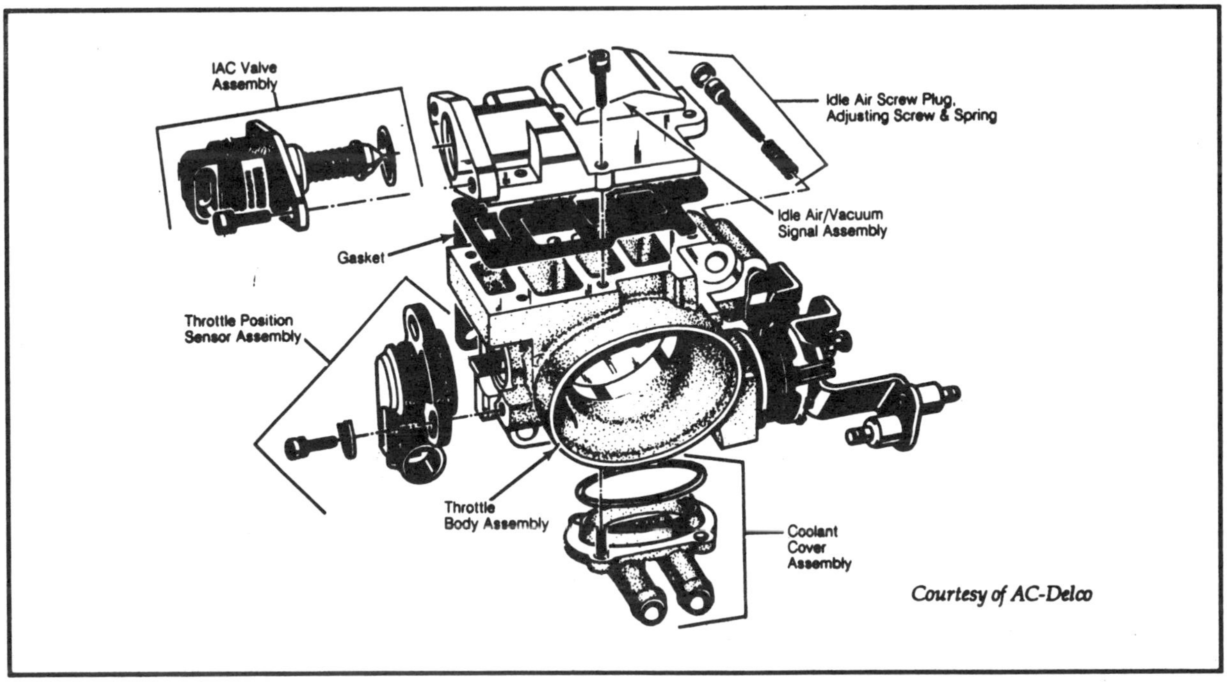

Typical 2.8L V6 Throttle Bodies

PFI Throttle Bodies (Cont.)

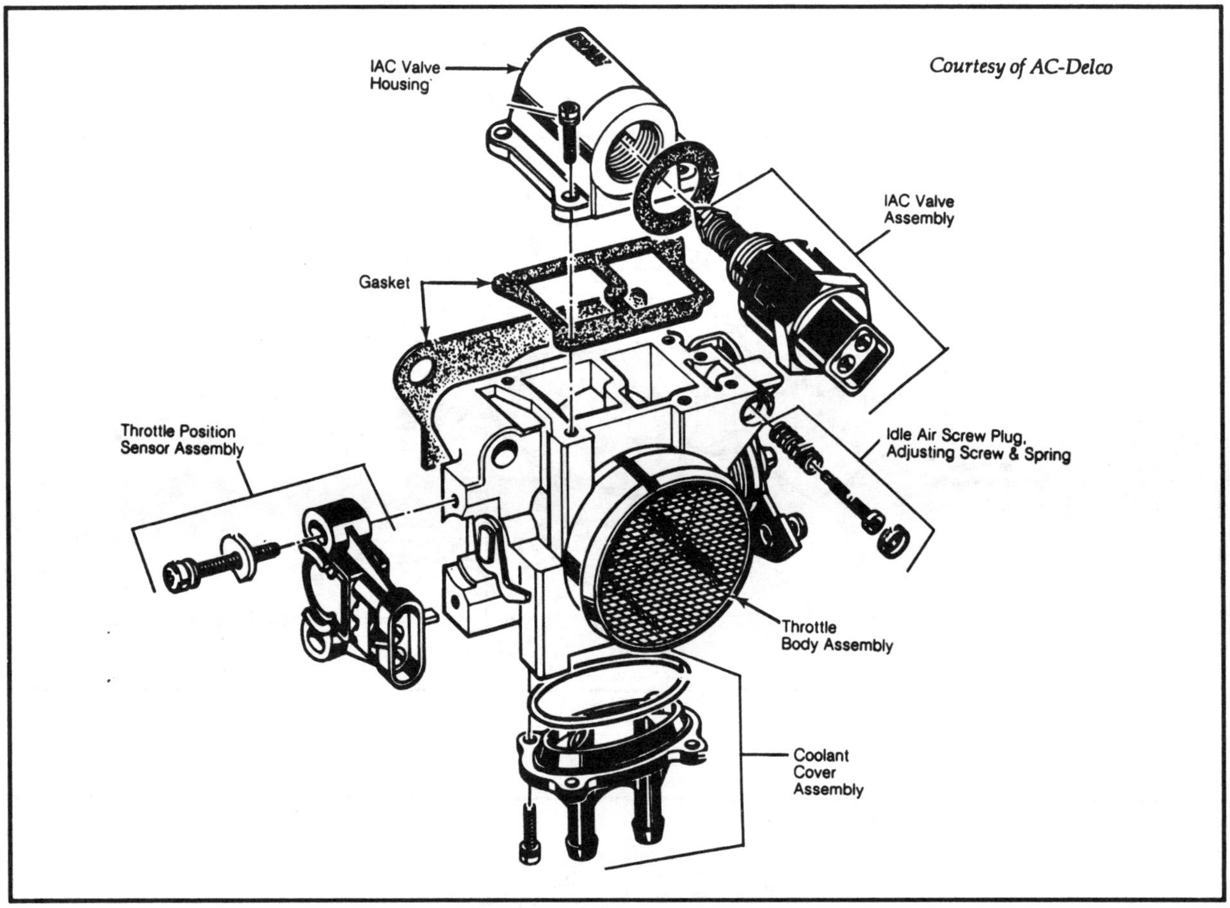

Typical 3.0L & 3.8L V6 Throttle Body

PFI Throttle Bodies (Cont.)

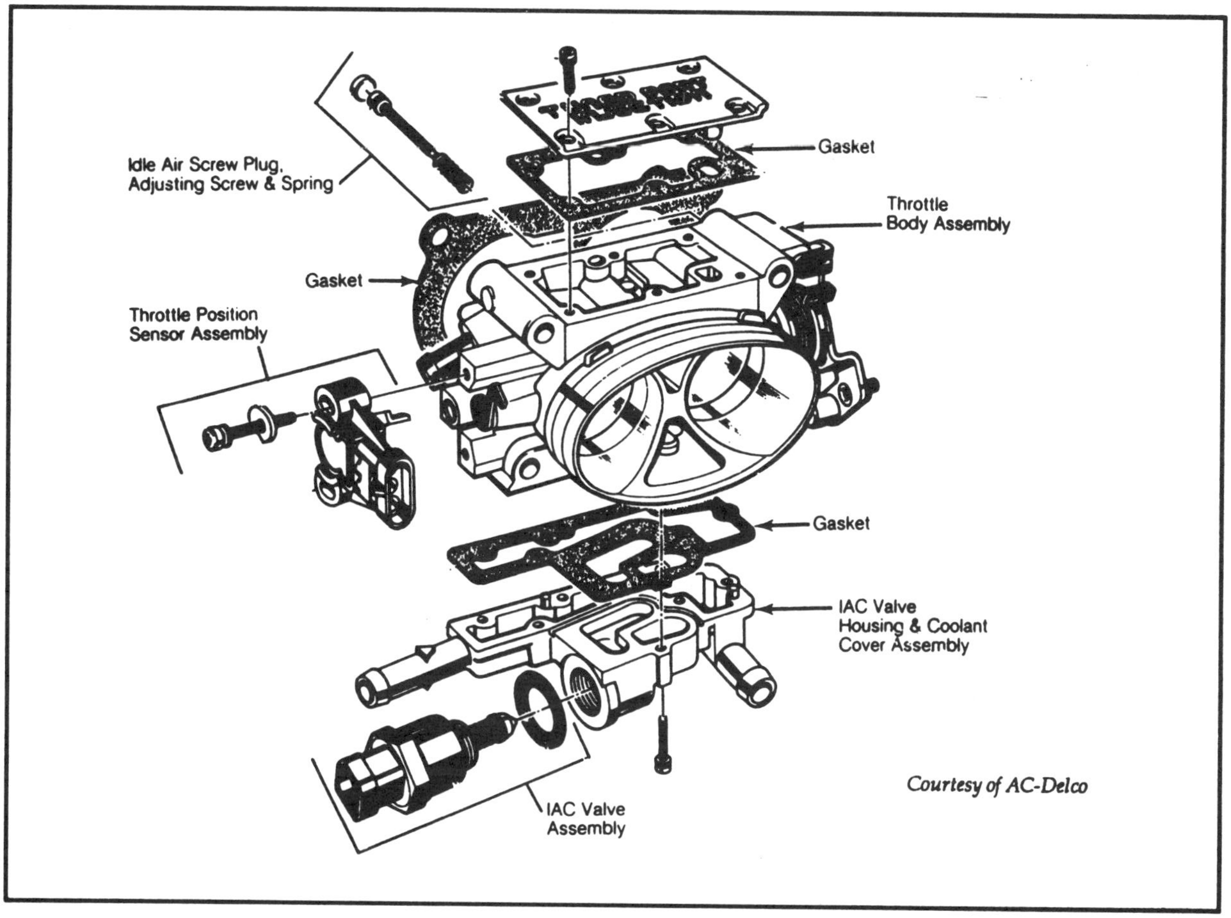

Typical 5.0L & 5.7L V8 Throttle Body

Using The Specifications

The number of adjustments which can be performed on the EFI equipped engines has been greatly reduced in comparison to the old carbureted systems of the past. In some cases, the following specifications will only be used to verify whether or not the ECM is receiving an "in range" signal. On some GM engines, the TPS is adjustable. On some it is not. If it is a non–adjustable TPS, and the signal line voltage is out of the window, then the TPS sensor may have to be replaced.

On almost all engines, even though the ECM controls the idle speed, the idle angle of the throttle plate is adjustable. The minimum idle speed adjustment will cure many stalling or rough idle complaints.

Use good common sense when applying these specification charts. Don't overlook the possibility of a circuit being just slightly out of range, but not causing a performance problem. Learn how the ECM logic searches for the correct inputs and actuator responses and keep in mind that all specs assume a fully-charged, operational battery is in use.

Specifications

TBI Engines

Application	TPS	Minimum Idle	Fuel Pressure	Volume in Pints Per 30 Seconds
TBI-220				
17086056	.45-.60 V[1]	500	9-13 psi	1
17086112,				
17086113	Less than l.25 V[1]	500[8]	9-13 psi	1
17086084,				
17086085	Less than 1.25 V[1]	500[8]	9-13 psi	1
17086086	.42-.54	675-725	9-13 psi	1
17086088	.42-.54	675-725	9-13 psi	1
17086090	Less than 1.0 V[1]	500-550[7]	9-13 psi	1
17087004	Less than 1.0 V[1]	500-550[7]	9-13 psi	1
17087004	Less than 1.0 V[1]	500-550[7]	9-13 psi	1
17807019	Less than 1.0 V[1]	500-550[7]	9-13 psi	1
17087021	Less than 1.0 V[1]	500-550[7]	9-13 psi	1
17087090	Less than 1.0 V[1]	500-550[7]	9-13 psi	1
17087195,				
17087196	Less than 1.0 V[1]	500-550[7]	9-13 psi	1
17087100,				
17087101	Less than 1.0 V[1]	500-550[7]	9-13 psi	1
TBI-200				
17081030 ('81-'83)	.50-.60 V	500-550[7]	9-13 psi	1
17081030 ('84)	.45-.55 V	500-550[7]	9-13 psi	1
17081031,				
17081032	.50-.60 V	500-550[7]	9-13 psi	1
17082030,				
17082031	.45-.55 V	500-550[7]	9-13 psi	1
17084034,				
17084036,				
17084035	.45-.55 V	500-550[7]	9-13 psi	1
17085056	.45-.60 V	550[2]	9-13 psi	1
17087030	.45-.55 V	550[2]	9-13 psi	1
TBI-300				
17085061 ('82 1.8L)	.50 V at idle	700[3]	9-13 psi	1
17085061	.5-1 V	775-825[4]	9-13 psi	1
17085062	.5-1 V	750-800[5]	9-13 psi	1
17085069	.5-1 V	750-800[5]	9-13 psi	1
17086063	.5-1 V	750-800[5]	9-13 psi	1
17085092,				
17085068	.50 V at idle			
17082060	.45-.60 V	500[5]	9-13 psi	1
17082061,				
17084060	.45-1.25 V	800[6]	9-13 psi	1
17082066	.57-2.07 V	700[3]	9-13 psi	1

Specifications

TBI Engines (Cont.)

Application	TPS	Minimum Idle	Fuel Pressure	Volume in Pints Per 30 Seconds
TBI 300 (Cont.)				
17083060,				
17084060,				
17084063,				
17084065	.45-1.25 V	500[6]	9-13 psi	1
17084066	.45-1.25	700	9-13 psi	1
17086109	.45-1.25 V	500[6]	9-13 psi	1
17085082	.45-1.25 V	625-675	9-13 psi	1
17085083	.45-1.25	625-675	9-13 psi	1
17086111	.45-1.25	500[6]	9-13 psi	1
17086063	.45-1.25	500[6]	9-13 psi	1
17086110	.45-1.25	500[6]	9-13 psi	1
17085108	.45-1.25 V	500[6]	9-13 psi	1
17085065	.45-1.25	500[6]	9-13 psi	1
TBI-400				
17082056,				
17082055,				
17083055	.45-60 V	475	9-13 psi	1
TBI-500				
17085083	.5-1 V	625-675	9-13 psi	1
17085082	.5-1 V	625-675	9-13 psi	1
TBI-700				
17087060,				
17087061	Less than 1.25 V[9]	575-625	9-13 psi	1
17087250	Less than 1.25 V[9]	575-625	9-13 psi	1
17087469	Less than 1.25 V[9]	575-625	9-13 psi	1
17087279,				
17087280	Less than 1.25 V[9]	575-625	9-13 psi	1
17087282,				
17087283	Less than 1.25 V[9]	625-675	9-13 psi	1
17087086,				
17087087	Less than 1.25 V[9]	625-675	9-13 psi	1

1 - Not adjustable.
2 - In drive.
3 - AT models. MT 800 RPM.
4 - MT/AT is 675-725 RPM on all models except Oldsmobile and Pontiacs. On Oldsmobile and Pontiac set minimum idle speed to 475-525 RPM.
5 - MT models. AT 475-525 RPM in Park.
6 - AT models. MT 775 RPM.
7 - AT models in Drive. MT models 600-650 RPM in Neutral.
8 - AT models in Park. MT models 775 RPM in Neutral.
9 - With ignition on, engine off.
A - Minimum idle speed on Beretta and Corsica should be 625-675 RPM.

Specifications

PFI Engines

Application	TPS	Minimum Idle	Fuel Pressure	Volume in Pints Per 30 Seconds
1984				
1.8L Turbo (vin J)	.50V at idle	675-725[1]	34-40	psi 1
3.8L (vin 3)	.35-.45V at idle	450-550[3]	34-40	psi 1
3.8L (vin 9)	.35-.45V at idle	450-550[3]	34-40	psi 1
1985				
1.8L Turbo (vin J)	[2,4]	[6]	34-40 psi	1
2.8L (vin S, W, 9)	.50-.60	MT 600-700	34-40 psi	1
		AT 500-600[3]	34-40 psi	1
3.0L (vin 3)	.50-.60	450-550[3]	34-40 psi	1
3.8L (vin 3)	.35-.44[6]	450-550[3]	34-40 psi	1
3.8L Turbo (vin 3)	.35-.44[6]	450-550[3]	34-40 psi	1
5.0L (vin F)	.465-.615	400[7]	34-40 psi	1
1986				
1.8L Turbo (vin J)	[2,4]	[6]	34-40 psi	1
2.8L (vin S, W, 9)	.50-.60	MT 600-700	40-47 psi	1
		AT 500-600[3]	40-47 psi	1
3.0L (vin L)	.50-.60	450-550[3]	37-43 psi	1
3.8L (vin B, 3)	.35-.44 6[4]	450-550[3]	37-43 psi	1
3.8L Turbo (vin 7)	.35-.44 6[4]	450-550[3]	37-43 psi	1
5.0L (vin F)	.465-.615	400[3]	40-47 psi	1
5.7L (vin 8)	.465-.615	400[3]	40-47 psi	1
1987				
2.0L Turbo (vin M)	.5	550-650[6]	35-48 psi	1
2.8L (vin S, W, 9)	.45-.65	MT 600-700	40-47 psi	1
		AT 500-600[3]	40-47 psi	1
3.0L (vin L)	.50-.59	450-550[3]	40-47 psi	1
3.8L (vin 3)	.35-.44[4]	450-550[3]	34-40 psi	1
3.8L Turbo (vin 7)	.35-.44[4]	450-550[3]	34-40 psi	1
5.0L (vin F)	.465-.615	400[3]	40-47 psi	1
5.8L (vin 8)	.465-.615	400[3]	40-47 psi	1
1988				
2.0L Turbo (vin M)	.5	550-650[6]	35-48 psi	1
2.3L (vin D)	.54	625-675	40-47 psi	1
2.8L (vin S, W, 9)	.55	500-600[3]	40-47 psi	1
3.0L (vin L)	.55	450-550[3]	31-42 psi	1
3.1L (vin T)	.55	450-550[3]	31-42 psi	1
3.3L (vin N)	.40	450-550[3]	40-44 psi	1
3.8L (vin 3)	.40	450-550[3]	34-40 psi	1
3.8L Turbo (vin 7)	.40	450-550[3]	40-47 psi	1
5.0L (vin F)	.54	450-550[3]	40-47 psi	1
5.7L (vin 8)	.48-.56	450-550[3]	40-47 psi	1

Specifications

PFI Engines (Cont.)

Application	TPS	Minimum Idle	Fuel Pressure	Volume in Pints Per 30 Seconds
1989				
2.0L Turbo (vin M)	.5	550-650[6]	35-48 psi	1
2.3L (vin A, D)	.54	625-675	40-47 psi	1
2.8L (vin S, W, 9)	.55	500-600[3]	40-47 psi	1
3.1L (vin T)	.55	450-550[3]	31-42 psi	1
3.3L (vin C)	.40	450-550[3]	34-40 psi	1
3.8L (vin C)	.40	450-550[3]	34-40 psi	1
5.7L (vin J)	.48-.56	450-550[3]	40-47 psi	1
1990				
2.0L Turbo (vin M)	.5	550-650[6]	35-48 psi	1
2.2L (vin G)	.5	550-650[6]	35-48 psi	1
2.3L (vin A, D)	.54	625-675	40-47 psi	1
2.8L (vin S, W, 9)	.55	500-600[3]	40-47 psi	1
3.1L (vin T)	.55	450-550[3]	31-42 psi	1
3.3L (vin C)	.40	450-550[3]	34-40 psi	1
3.8L (vin C)	.40	450-550[3]	34-40 psi	1
5.7L (vin J)	.48-.56	450-550[3]	40-47 psi	1

1 - In Park. On MT, set to 775-825 RPM.
2 - Not Adjustable.
3 - AT in Drive.
4 - Maximum voltage value of .44 V.
5 - Less than 1.25 V with throttle closed at idle or with ignition on.
6 - Minimum air rate is reset by ECM when vehicle is driven at about 35 MPH
7 - In Drive. On MT, set to 450 RPM.
8 - AT in Neutral.
9 - No RPM setting. IAC acceptable value is 5-45 "counts" on scan tool.
10 - No RPM setting. Minimum air rate will be reset by ECM when ignition is turned on, then off.

Review of Key Concepts

- **Diagnosing and repairing GM EFI engines requires a logical sequence diagnostic procedure.**

- **This requires professional tools, data sources and training.**

- **This manual provides the vital information pertaining to GM.**

- **GM has evolved the current EFI systems throughout the Eighties to date.**

Table of Contents

Section II

Section II Outline

I. Logical and Diagnostic Procedures

A. Set up shop procedure to follow a logical diagnostic sequence.
   1. Identify the customer complaint(s) first.
   2. Perform a complete visual inspection of the vehicle.
   3. Carefully identify the vehicle and the systems which are used on the engine to be tested.
   4. Test and verify the basic engine systems.
   5. Enter the self–diagnostic system and retrieve any fault codes using the Diagnostic Circuit Check procedure.
      a. Retrieve all stored codes.
      b. Clear the codes and then operate the vehicle to see if the codes return.
      c. Repair hard code failures.
      d. Test each management system group to verify that the system is operational.

B. Create and utilize a comprehensive diagnostic report to use as a guide for diagnosis.
   1. Compose a customized report for your shop.
   2. Compose a special report for GM EFI systems.
   3. Add any other areas of reporting which fit your shop capabilities.
   4. The completed diagnostic report form can be used as a sales recommendation sheet for your service advisor.

C. The technician should understand basic EFI operating principles.
   1. GM fuel management theory and application.
   2. GM electronic timing management theory and application.
   3. GM idle management system operation.

D. GM ignition systems evolution for EFI engines.
   1. GM HEI
   2. GM HEI/EST
   3. GM HEI/EST/ESC
   4. GM DIS

Logical Diagnostic Sequence

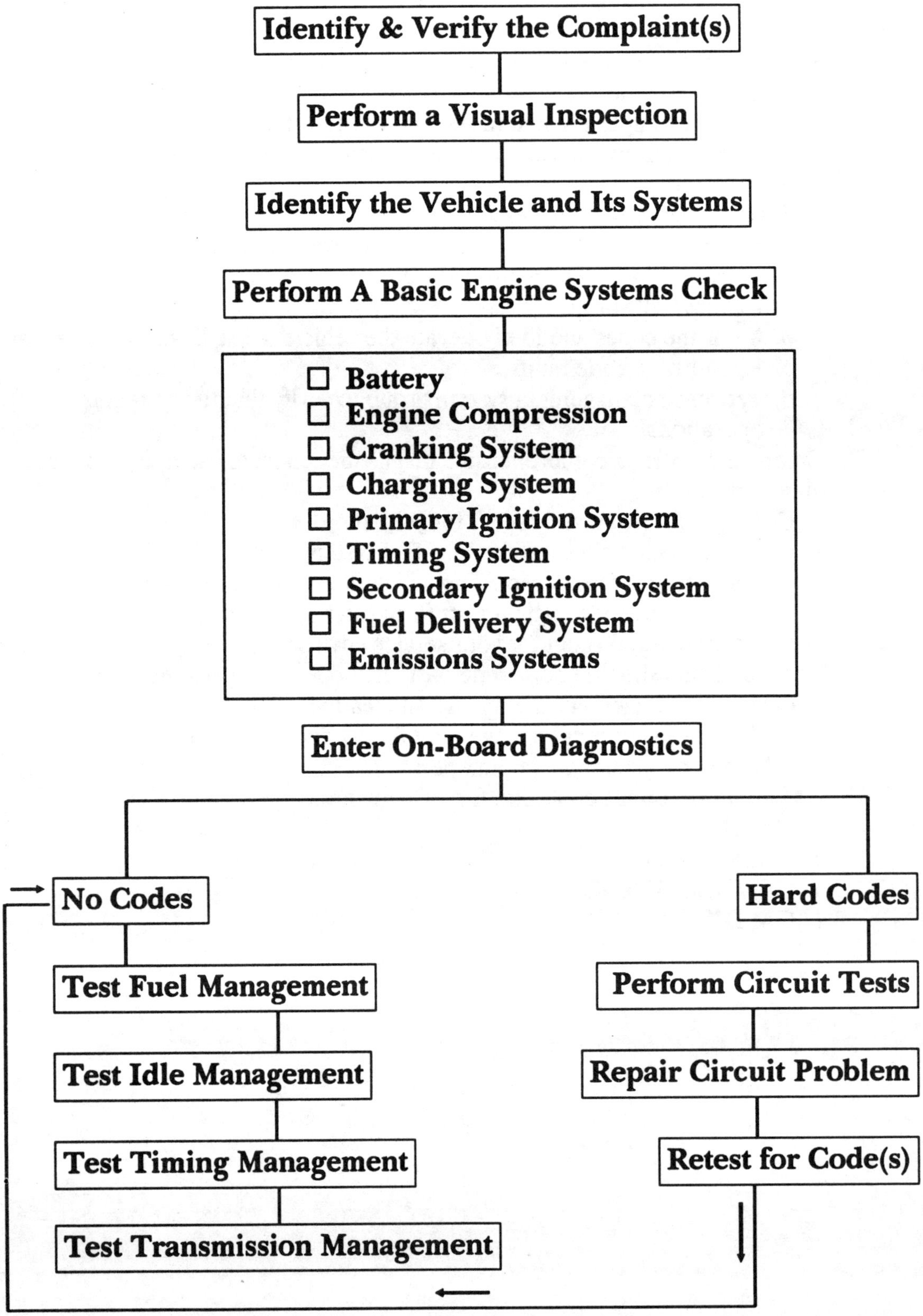

Diagnostic Report

Customer: ______________________ Vehicle Year & Make: __________________

Address: ______________________ Model: __________________

______________________ Mileage: __________________

Phone: ( )__________________

Driveability Complaint(s): ____

____

____

VISUAL INSPECTION

	OK	Needs Attention
Battery Cables	☐	☐
Air Filter	☐	☐
Oil Level	☐	☐
Oil Condition	☐	☐
Oil Leaks	☐	☐
Belts	☐	☐
Hoses	☐	☐
Coolant	☐	☐
Transmission	☐	☐
Emission Devices	☐	☐
Tires	☐	☐
Shocks	☐	☐

ENGINE TESTS

	OK	Needs Attention
Battery	☐	☐
Cranking System	☐	☐
Compression	☐	☐
Charging System	☐	☐
Ignition System	☐	☐
Timing System	☐	☐
Fuel System	☐	☐

On Board Computer System Diagnostic Report

Customer: ______________

	Yes	No
Fault Codes	☐	☐

____

____

Fuel Pressure ______________

Fuel Flow ______________

Fuel Quality ______________

FUEL MANAGEMENT:

Loop Status ______________

Coolant Temp ______________

Air Temp ______________

Throttle Body Temp ______________

Throttle Position ______________

MAP ______________

BARO ______________

RPM ______________

Oxygen Volts idle ______ to ______

cruise ______ to ______

load ______ to ______

MAF

idle cruise load

Air Flow

idle cruise load

Cross Counts

idle cruise load

Integrator

idle cruise load

IDLE MANAGEMENT:

No Load ______________

Load Applied ______________

Trans. Applied ______________

AC "ON" ______________

P.S. Applied ______________

Vehicle Year & Make: ______________

Vehicle Engine: ______________

ROAD TEST

FUEL MANAGEMENT

MAF

idle cruise load

Oxygen V idle ______ to ______

cruise ______ to ______

accelerate ______________

decelerate ______________

Inj. Pulse idle ______ to ______

cruise ______ to ______

accelerate ______________

decelerate ______________

Cross Counts

idle cruise load

Integrator

idle cruise load

Block Learn

idle cruise load

Timing idle ______ to ______

cruise ______ to ______

accelerate ______________

decelerate ______________

TRANSMISSION MANAGEMENT

TCC ☐ OK ☐ Not OK

Gear Switches ______________

VSS ______________

P/N ______________

Injector Cleaning or Intake System

____

Cruise Control ☐ OK ☐ Not OK

GM Fuel Management

GM uses different fuel management system types on different engine platforms. It is important for the technician to determine which management system is being utilized. The throttle body engines all use the speed density method of computing engine fuel requirements. The Pontiac Fiero was the only port fuel engine to use this method initially. The speed density method utilizes the information from the RPM and MAP signals to compute fuel injector pulse widths. The ECM corrects those computations for temperature and oxygen voltage.

The other GM port fuel engines use the air flow controlled type of system. This method is called mass airflow because the computer receives a signal from an airflow sensor which informs the computer of the volume and mass of the air intake of the engine. All of the early GM engines used Delco mass airflow sensors except the Corvette, the IROC–Z car and the Pontiac Trans Am. Those engines used a hot wire type of mass airflow sensor which is manufactured by Bosch. When Buick redesigned the 3.0L V6 and the 3.8L V6 engines and created the 3300 and the 3800, they began using a new style mass airflow sensor built by Hitachi. On any of these systems, the important thing to remember is that the computer issues injector pulse width commands primarily based on engine fuel requirements determined principally from the airflow sensor and RPM. The computer corrects these computations for temperature variances and oxygen voltage.

In 1989, GM's Chevrolet–Pontiac–Canada Group produced engines, dropped the use of mass airflow sensors and started using MAP sensors with the computer being programmed for speed–density calculations. Actually, both methods work quite well and the efficiency of either system is directly proportionate to the accuracy of the main input sensors or the programming done within the computer itself.

Each operating engine mode has computed fuel injector pulse widths and frequencies stored in the computer for the optimum engine power along with acceptable fuel economy and emissions.

The technician must develop the discipline of verifying the fuel management system in idle, cruise and engine-loaded modes, contingent upon engine and air intake temperature. The oxygen sensor acts like an on-board exhaust gas analyzer by responding within certain "predictable" ranges if the computer is administering the proper fuel pulse width control.

GM Fuel Management

Cont.

On a cold engine, the air/fuel ratio should be very rich. The oxygen sensor cannot be counted on for information in this mode because the sensor only produces the properly corresponding voltage when it is at operating temperatures. Some GM engines use an electrically-heated sensor which should reach the desired operating temperature quicker than the devices heated only by the engine exhaust. But the best indicator of fuel enrichment on a cold engine is the injector pulse width. When the engine is cold, the injector pulse width should be well above 3 milliseconds. If the engine intake air is also cold, this increases the pulse width even more.

On a warm engine (with coolant above 190° F), the oxygen sensor should be active enough to verify whether the computer can manage the fuel system. If the system goes into a closed loop, the oxygen voltage rapidly switches back and forth above and below 500 millivolts. This can also be verified on some scanners as a rapid alternating flashing of the Rich-Lean indicator LEDs.

The technician should increase the engine RPM to above 2500 RPMs and note whether or not the oxygen sensor still indicates closed loop by continuing to switch above and below the 500 millivolt range. Next, he should place the vehicle in gear and road test the vehicle at idle, slow cruise and high speed cruise. The oxygen voltage should increase immediately when the vehicle is accelerated and begin switching again when the cruise level is resumed. On some vehicles, when they enter the deceleration mode, the computer significantly reduces the amount of fuel delivered in order to conserve fuel. This may result in a momentarily low oxygen voltage but does not always occur on GM vehicles.

If in any of these operating modes the fuel management system appears to be out of control, the oxygen voltage may be stuck high if the system is unable to reduce the fuel for proper enleanment. The oxygen voltage may be stuck low if the system is unable to increase the fuel for proper enrichment.If this occurs, the technician should first verify that the oxygen sensor can respond to enrichment or enleanment, thereby verifying system fuel control.

If the oxygen voltage is stuck high, then produce an abundance of exhaust oxygen by opening a large vacuum leak to see if the voltage will come down. If the sensor is capable of "seeing" oxygen, the sensor will rapidly drop its voltage to below 100 millivolts. Judging the size of the vacuum leak required to accomplish this lower voltage will also indicate the amount of enrichment to which the engine is being subjected. If the oxygen voltage is stuck low, the technician should create engine enrichment by reducing the intake air to the engine. An alternative method would be to bleed off the vacuum to the MAP sensor. Low vacuum to the MAP sensor should cause the computer to significantly increase the injector pulse width which should cause the oxygen voltage to increase rapidly above 900 millivolts. If rich or

GM Fuel Management

Cont.

lean system emulation causes the oxygen sensor to respond, this verifies that the oxygen sensor is capable of detecting and indicating air fuel ratios above or below stochiometric. If not, replace the oxygen sensor.

If the system continues to run rich, keep in mind that the factors that are involved concern possible causes of abundant fuel or an absence of air. Restricted air intake, high fuel pressure, MAF frequency, MAP voltage, RPM, CTS, charge temperature and throttle position voltage are the possible causes. Don't forget that the canister purge system can also cause enrichment symptoms. Fuel-contaminated engine oil can also cause the oxygen voltage to stay high.

If the system continues to run lean, keep in mind that the factors involved here are possible causes of abundant air or the absence of fuel. Vacuum leaks, low fuel pressure or volume, MAF frequency, MAP voltage, RPM, CTS, charge temperature and throttle position voltage are some of the possible causes. Don't forget that the oxygen sensor vent could be plugged or contaminated with exhaust gas from a manifold or exhaust system leak. On the bigger engines, the air system could pump air upstream ahead of the oxygen sensor.

Oxygen voltage information can be observed through the scanner. If the serial communicator is inoperative, the technician can tap directly into the oxygen sensor circuit with a high impedance voltmeter and observe the voltage amounts and rate of change directly.

Oxygen Sensor Voltage

Oxygen Sensor Graph

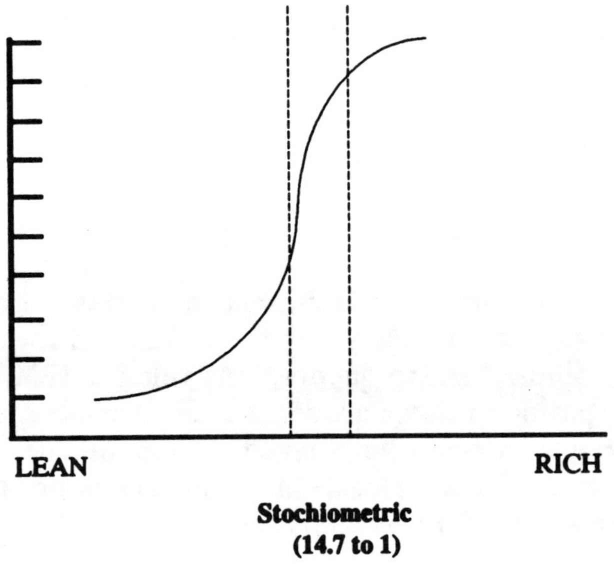

Air/Fuel Mixture

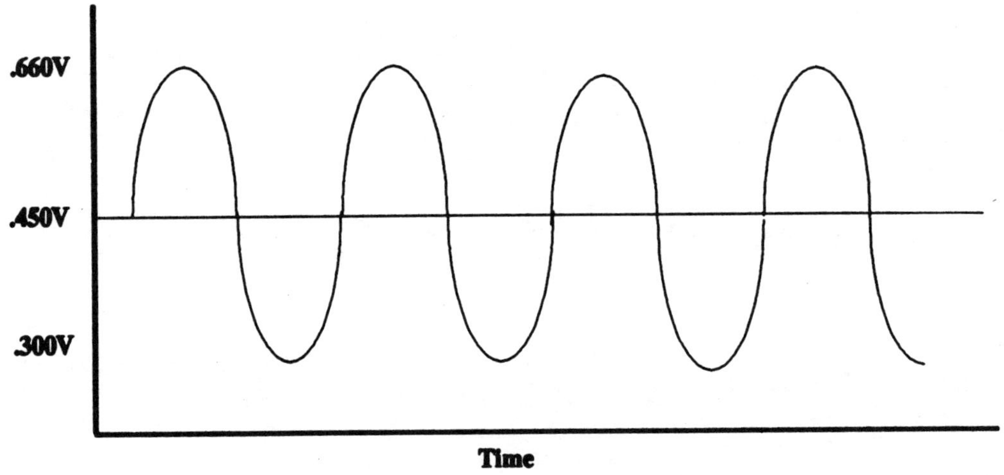

Rapid voltage production above and below reference indicates an active sensor and good computer control response.

Oxygen Sensor Voltage (Cont.)

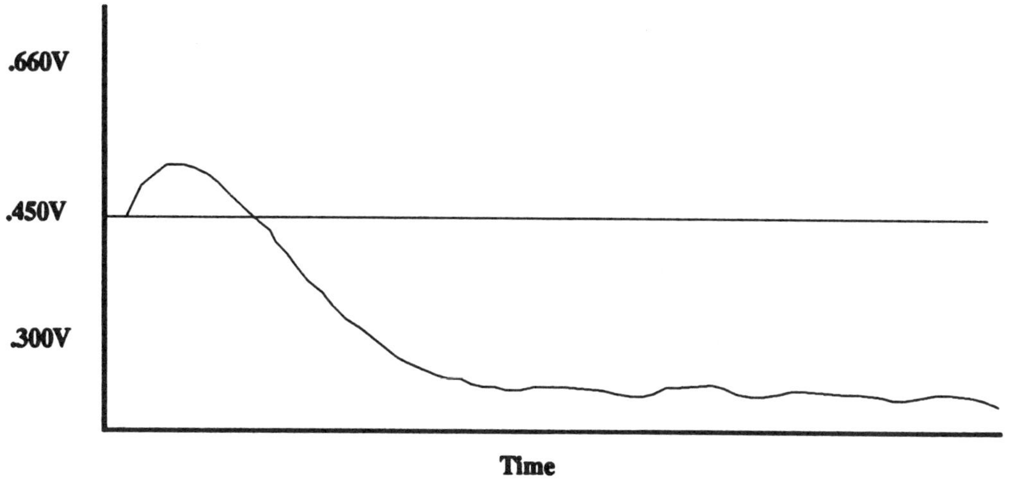

Oxygen voltage below reference too long = abundant oxygen; will result in a code 44.

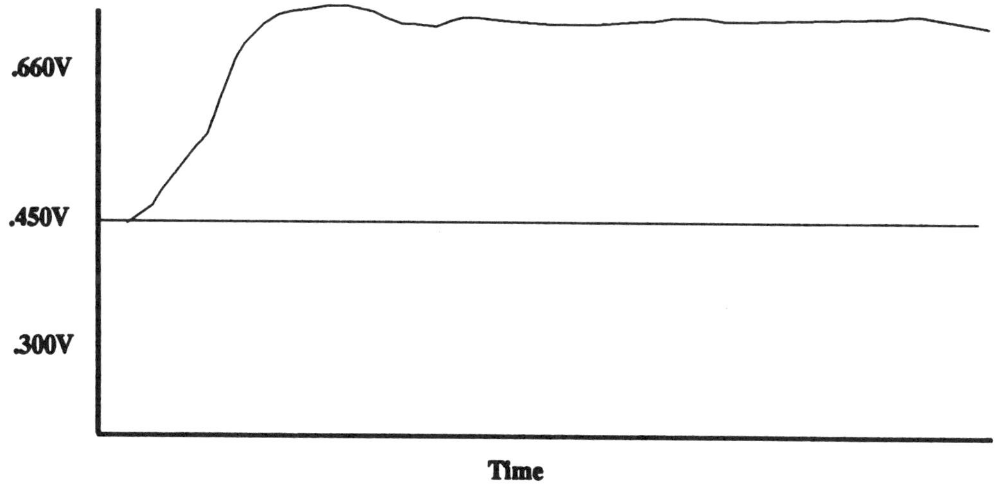

Oxygen voltage above reference too long = absence of oxygen; will result in a code 45.

Oxygen Sensor Voltage (Cont.)

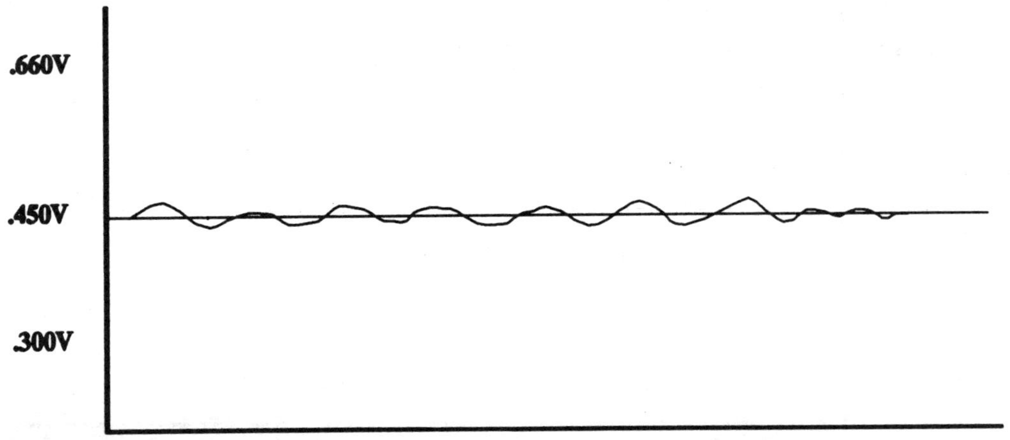

Oxygen voltage at reference too long – no switching = no oxygen indication; will result in a code 13.

GM Fuel Management

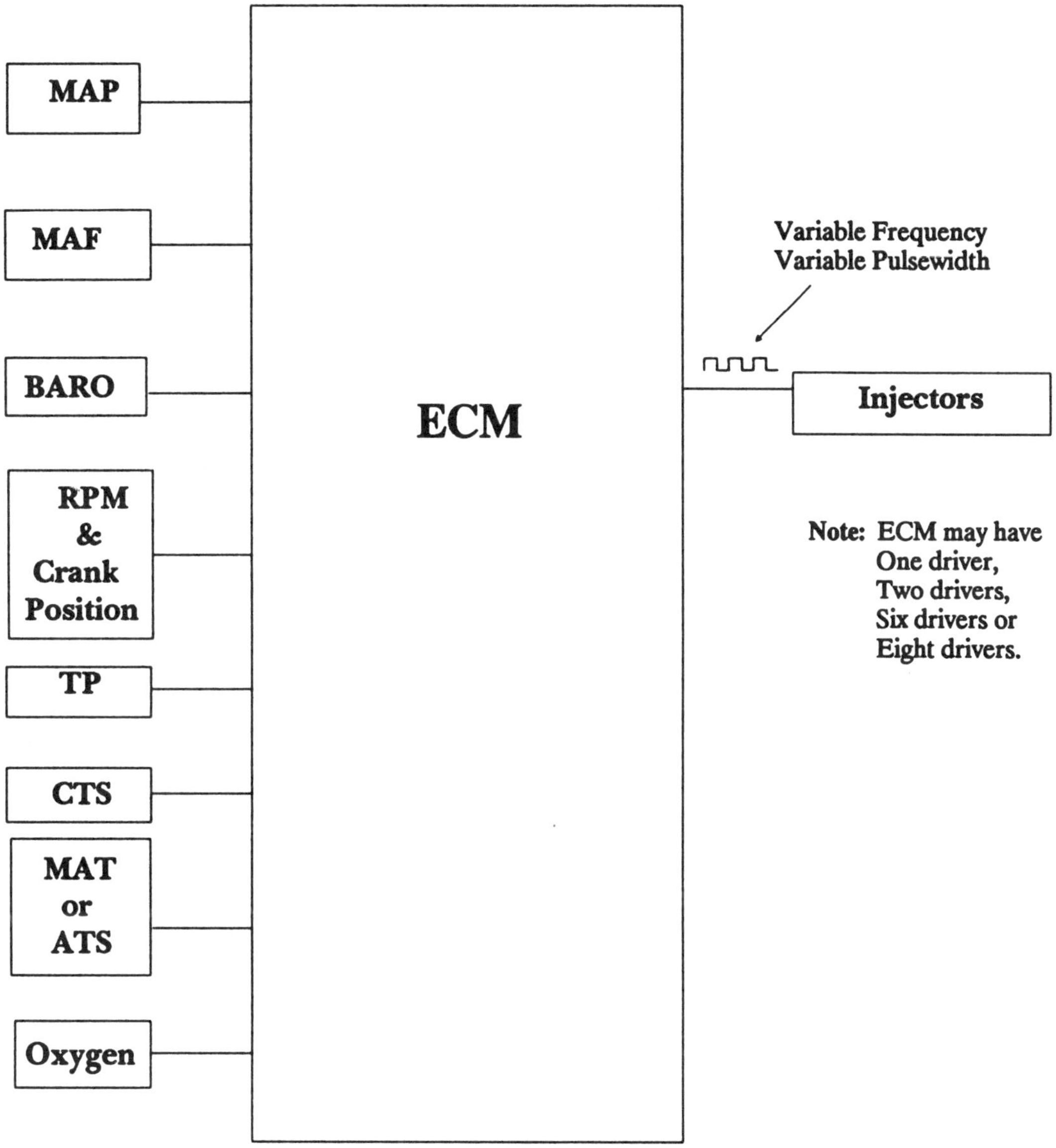

On Speed-Density System, the main pulsewidth control input signals are MAP and RPM. The other inputs are for pulsewidth correction.

On Air Flow Controlled System, the main pulsewidth control input signals are MAF and RPM. The other inputs are, again, used for pulsewidth correction.

GM Timing Management

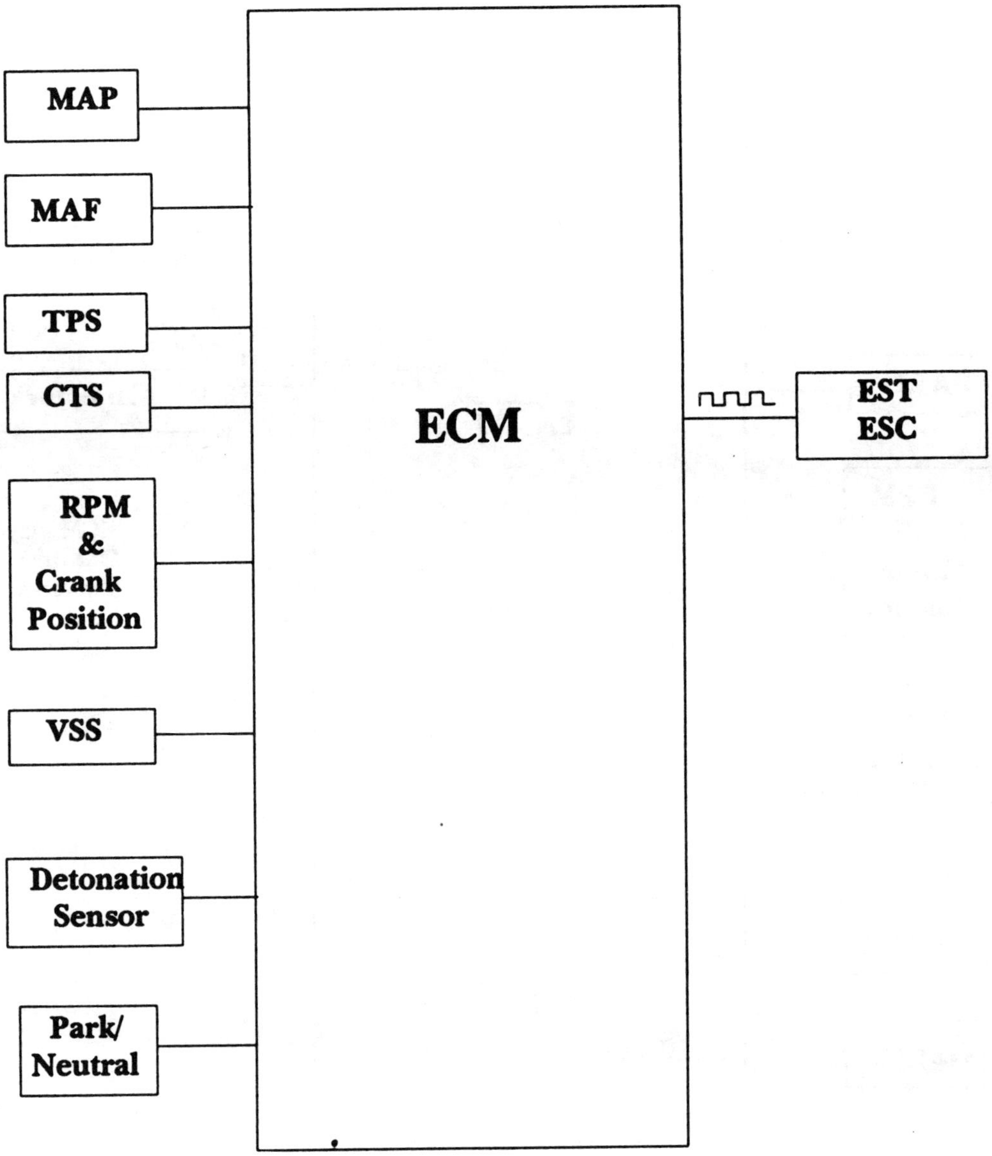

These sensors represent the input information used by the ECM to make timing management decisions

GM Idle Management

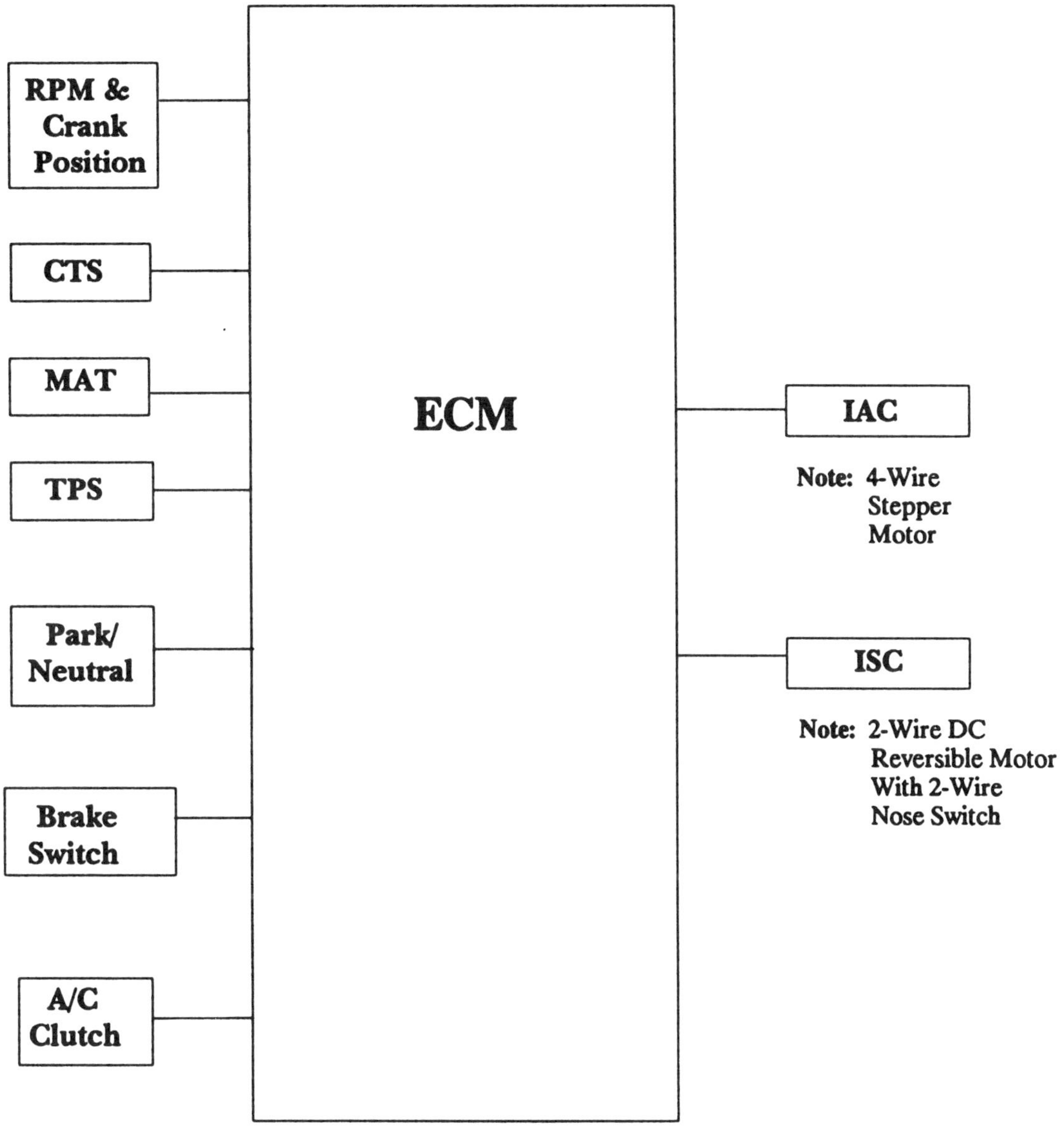

These sensors represent the input information used by the ECM to make idle management decisions.

GM Ignition System Evolution

All GM Fuel Injected Engines use some form of High Energy Ignition or HEI. Even the new distributorless ignition systems are high energy. Since the distributor reference signal is the only signal without which the ECM cannot operate the injectors signal, it would serve the technician well to become familiar with how that signal originates with each form of HEI.

In 1975, HEI was introduced on GM vehicles. The trigger signal which causes the power transistor to switch the primary current ON and OFF, inducing high voltage in the secondary, is generated from a wire-wound coil called the coil pickup. The output voltage generated is AC. The signal signature looks something like this.

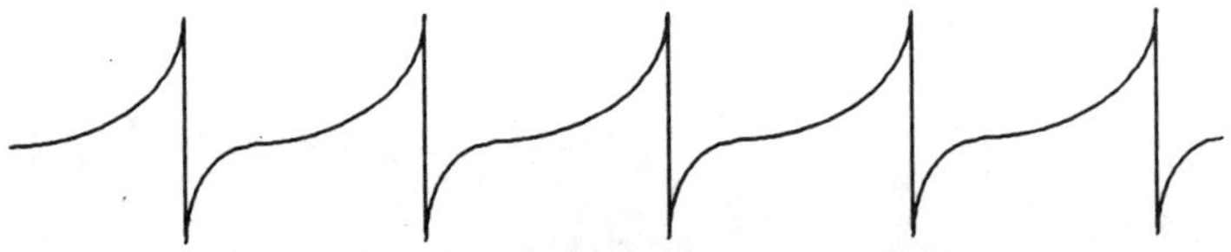

The ECM controlled timing management systems use an advanced form of HEI called HEI/EST. This system uses a seven pin ignition module which interfaces with the ECM to facilitate timing control. The schematic below is a typical circuit used for HEI/EST.

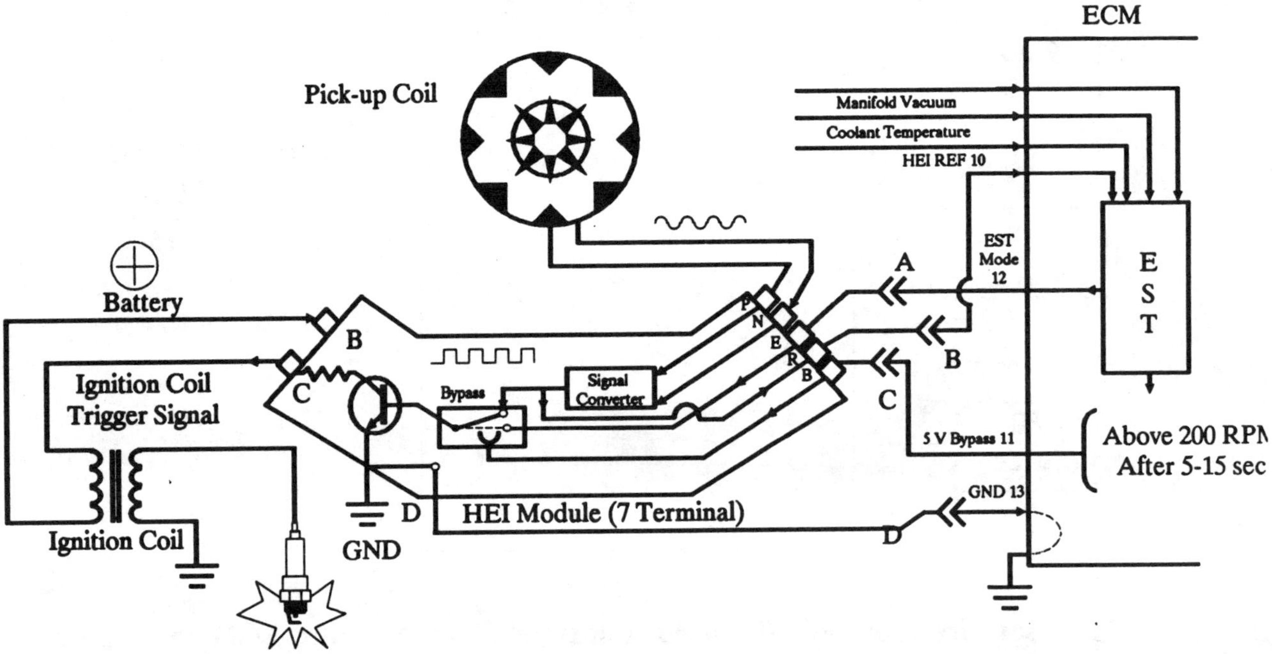

GM Ignition System Evolution

Cont.

Cranking or Module Mode of EST Operation

During an engine crank, the ECM does not put 5 volts on the bypass line. This allows the reference signal to be converted to digital and applied directly to the module power transistor. This means the timing is controlled by the ignition module.

Bypass or Running Mode of EST Operation

When the engine RPM indicates to the ECM that the engine has started, then the ECM puts 5 volts on the bypass line which switches the gate and allows the ECM to control the timing of the transistor switching process via the EST line. The reference signal is generated as AC voltage and converted within the ignition module to a DC digital pulse. This pulse is used by the ECM to trigger the injector firing pulses. The time shifted EST signal is applied to the power transistor during this mode. Therefore, the module is bypassed and timing is controlled by the ECM. Problems with either the EST or the bypass line will set a code 42. If the EST line is grounded, the engine will not start.

Hall Effect

There is a variation from the conventional HEI/EST distributor. Some GM engines use a Hall Effect Switch mounted in the distributor above the coil pickup. This switch is used to generate a separate DC digital pulse which is fed directly into the ECM. The reference terminal is left blank. The Hall Switch signal is used by the computer to fire the injector pulses and the EST section time shifts that signal for use by the ECM and the ignition module as the time modified EST signal.

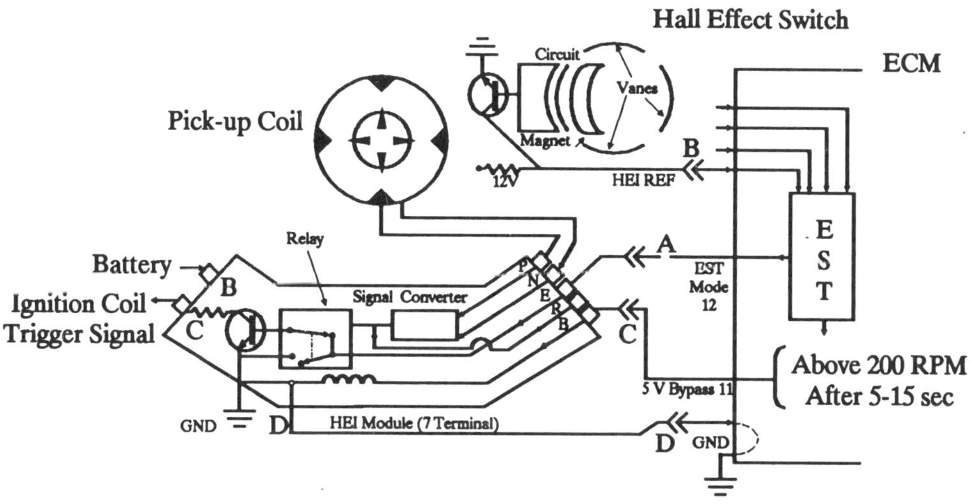

GM Ignition System Evolution

Cont.

Electronic Spark Control

The EST system described in the previous articles is the ECM controlled method of controlling timing advance. The EST subsection of the ECM is designed to replace the old mechanical advance controls and it does it quite well. The amount of timing advance and the speed at which advance control occurs has allowed GM to significantly increase the engine performance characteristics through this more sophisticated control. The problem of controlling spark knock in some engines had been addressed in the 70's by the use of a special retard system called Electronic Spark Control or ESC. The original ESC controller was a small metal box module which was used on the pickups and vans which could electronically retard the spark when detonation was detected. This system was adapted to interface with the ECM's EST timing system in 1981. This adapted version was used on the passenger car engines which might also encounter premature engine failure due to detonation. The engines which used this system were usually the turbocharged engines. The early ESC module was refined in the early 80's and became a five pin plastic module which interfaces more simplistically with the ECM.The illustration below shows the ESC arrangement used to date on many GM engines.

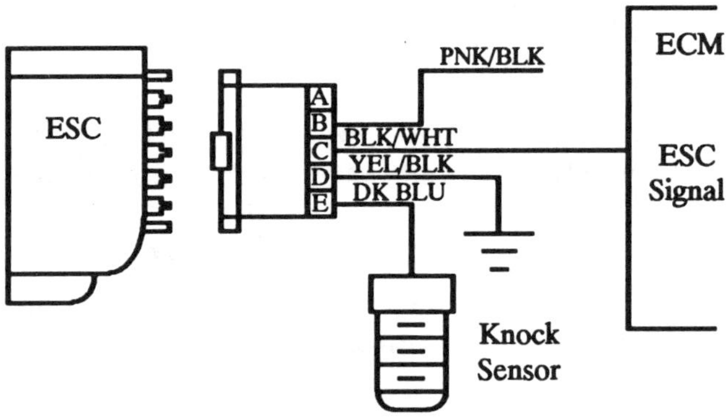

When this system is operating properly, the ESC module supplies a voltage (8-10 volts) on the ESC signal line to the ECM when no spark knock is detected. If the knock sensor generates a signal which the ESC module detects as engine knock, then the ESC module drops the signal line voltage low and the ECM then retards the spark timing by modifying the EST timing control signal to the ignition module. The ECM is programmed to expect the signal line voltage to be high under normal operation. If the signal line voltage goes low, the ECM expects the timing retard change which it implements to correct the spark knock. If the signal line voltage does not return to its normal high state, the ECM interprets that condition as a problem with the ESC system and the ECM then turns "ON" the Check Engine Lamp and stores a code 43. On the early ESC system, the detonation sensor could be disconnected and the ESC system would simply not operate. On the later version

GM Ignition System Evolution

Cont.

of this system, disconnecting the knock sensor will cause the ESC system to continually set a code 43.

The technician can test the operation of the ESC system by road testing the vehicle in the Field Service Mode. When a hard acceleration is applied, the ECM should detect detonation. If it does, the ECM will flash the Check Engine Lamp 5 times per second for 3 seconds, indicating that the detonation sensor signalled that detonation occurred and the ESC controller informed the ECM that timing should be retarded. If the ESC system does not operate properly, test the ESC module power and ground circuits, then test the knock sensor to verify that it can generate an AC voltage when knock is generated. Then test the ESC signal line circuit for opens or grounds. Check for possible signal interference on the signal line such as spark plug wire, etc.

The latest version of ESC used on many GM engines features the elimination of the external ESC module. The ESC module control circuitry is located in the MEM-CAL chip. This system has the knock sensor sending its AC signal directly into the ECM via the knock signal wire. The ESC portion of the MEM-CAL sends a signal to the EST controlled subsection if knock is detected. This is how the ECM controls the timing retard in the event that spark knock is detected. The illustration below shows the simplified wiring of this new ESC system.

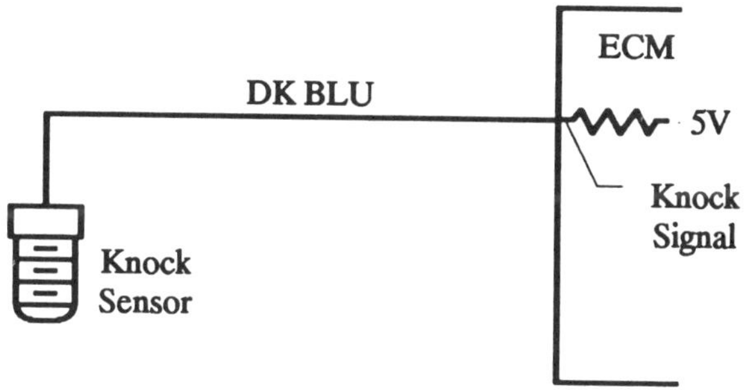

The circuitry within the knock sensor pulls the ECM supplied 5 volts to approximately 2.5 volts. The AC signal generated by the knock sensor will cause that signal line voltage level to vary when knock is detected. The amplitude and frequency of the AC signal will vary depending upon the knock level. An inoperative knock sensor, a shorted or open knock signal line or an inoperative ESC section within the MEM-CAL can cause a code 43 on this system. If the ESC circuitry and devices are working, and the ECM is responding to signal changes by retarding the timing when knock is detected, then the timing should be retarded whenever knock is produced. There is a serial data stream word for knock retard. The scanner should display the degrees of retard if retard is present.

GM Ignition System Evolution

Cont.

In 1984, GM introduced distributorless ignition on the 3.8L Buick engine. The new system was called Computer Controlled Coil Induction or CCCI (C^3I). The first version of C^3I was called Type I. The coils were molded together in one coil pack. The reference signal was generated directly from the crankshaft sensor. The synchronization signal was generated by a Hall Effect camshaft sensor. It was mounted on a small distributor shaft through the former distributor mounting hole in the manifold. The ECM uses the cam signal to determine which of the three coils to fire at the right time. The engine would not start with either of these signals missing. The illustration below is a typical 3.8L engine with C^3I.

Courtesy of AC-Delco

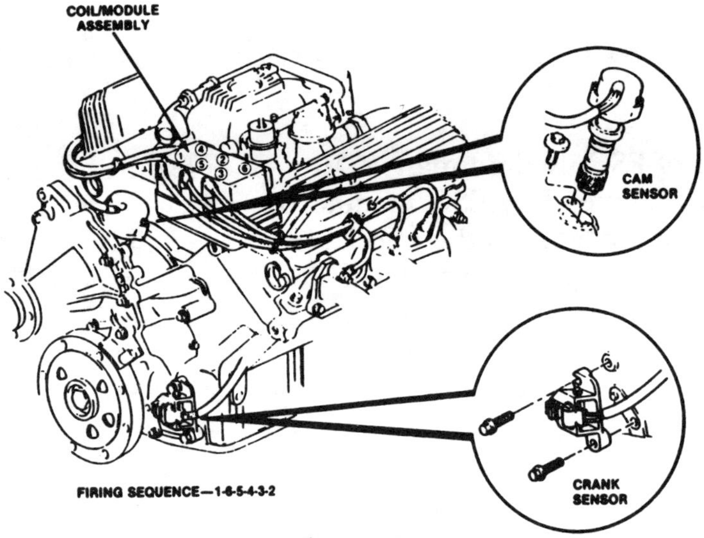

Buick then released Type II C^3I which featured individual coils which could be replaced individually. The ignition module used on Type II was of a different design also. The camshaft sensor was relocated to the engine timing cover. The 3.0L V6 was also converted to distributorless ignition with its own version of C^3I. The 3.0L is not a sequential fuel injection engine, so it does not require a cam signal for injector synchronization. However, the ignition module needs a synchronization signal in order to determine coil firing sequence, so the input sensor was turned into a dual Hall Effect sensor mounted on the crankshaft balancer which has two sets of metal vanes interrupting two Hall Effect sensor fields. One half of the sensor assembly signals crank position and engine RPM. The other half signals cylinder position or synchronization.

GM Ignition System Evolution

Cont.

The 3.8L V6 was upgraded to the new high-tech 3800 which featured a new version of C^3I called Type III Fast Start. The Fast Start version has a dual Hall Effect crankshaft sensor interrupted by two sets of metal vanes on the crankshaft balancer. One signal changes state three times per revolution and the other high resolution signal changes state 18 times per revolution. Since the 3800 is scquentially fuel injected, there is also a camshaft sensor located on the engine timing cover.

The common feature which binds all of these Buick distributorless systems together is the 14 pin connector on the module. All of the C^3I systems are triggered and synchronized by Hall Effect sensors. The illustration below is a typical 3800 V6 engine with Fast Start C^3I.

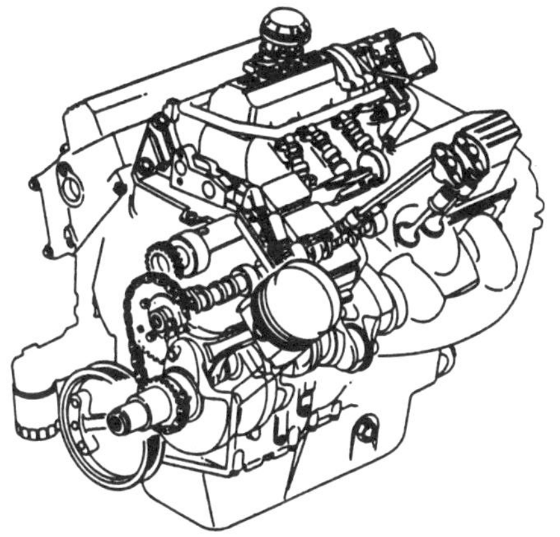

Courtesy of AC-Delco

The 3.0L V6 has also been upgraded to the new 3300 V6. The Type I coil packs are not interchangeable with any other GM division system. The Type II individual coils are interchangeable with other GM DIS systems.

In 1987, the Chevrolet-Pontiac-Canada group jumped onto the distributorless system bandwagon by releasing their own system called Direct Ignition System or DIS. The DIS system is used on the 2.0L L4, the 2.5L L4, the 2.8L V6 and the 3.1L V6. DIS does not require a cam signal because these engines are all non-sequential. The module does need a synchronization signal in order to make proper coil firing selection. The crankshaft signal is generated by a wire-wound coil pickup which is triggered by a special reluctor wheel made onto the engine crankshaft. The slots or notches on that reluctor are placed in a manner which allows the generated signal to have one of the AC voltage-state changes inform the ignition module of the cylinder position. This is how the module makes coil selection. The DIS coils are individual coils which are interchangeable with each other, or with Type II coils from C^3I. The coils have been failing due to the high secondary voltage which

GM Ignition System Evolution

Cont.

has been firing through the insulation to ground. Visually inspect the coils thoroughly to detect this condition. The illustration below represents a typical DIS application.

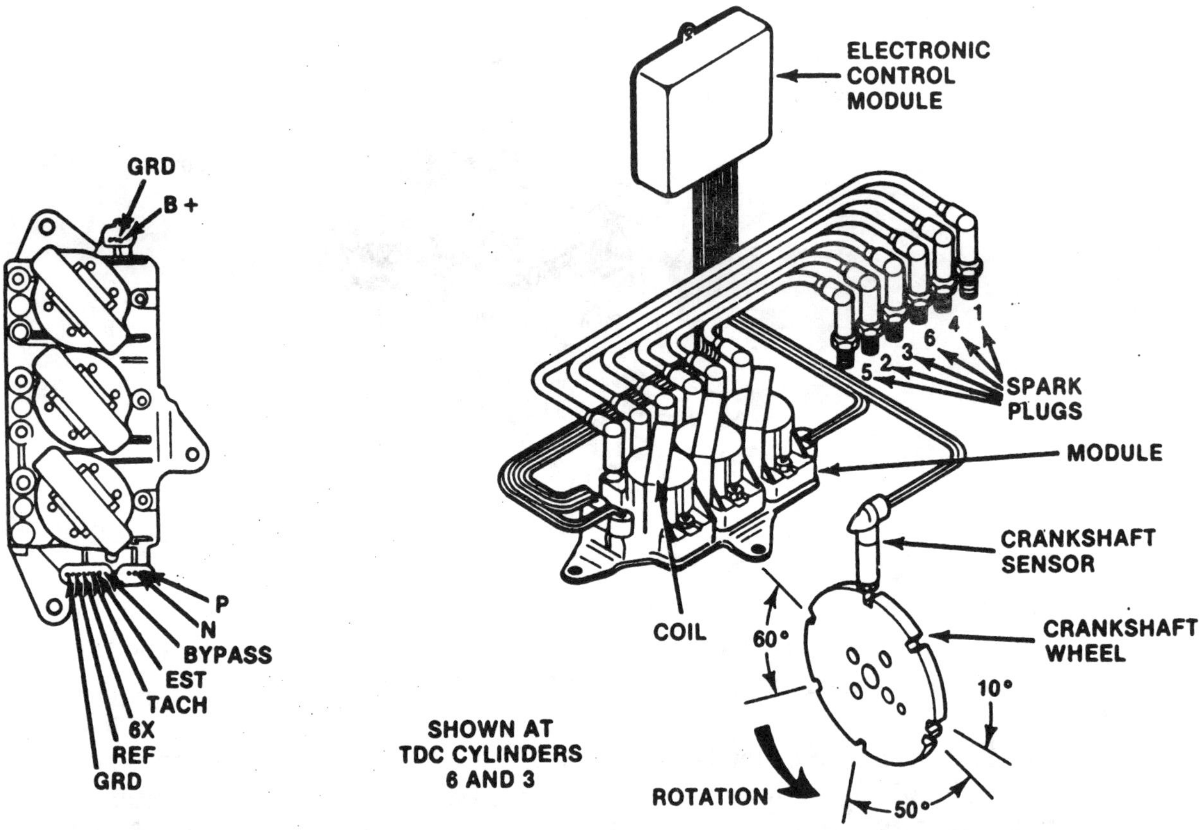

Courtesy of AC-Delco

In 1988, Oldsmobile released the new Quad 4 engine. This unique engine featured another new distributorless ignition system called Integrated Direct Ignition or IDI. The whole ignition system with the exception of the crankshaft sensor was located seemingly inside the engine underneath the top engine cover. The trigger

GM Ignition System Evolution

Cont.

signal, synch pulse included, is generated from a notched reluctor wheel inducing AC voltage signals from the wire-wound coil pickup sensor. The coils and the ignition module are located in a special shell housing underneath the cover and they are not interchangeable with any other GM DIS system. There are no spark plug wires. The coils discharge the high voltage directly into some copper strips which are routed to the spark plugs. The illustration below is a typical IDI application.

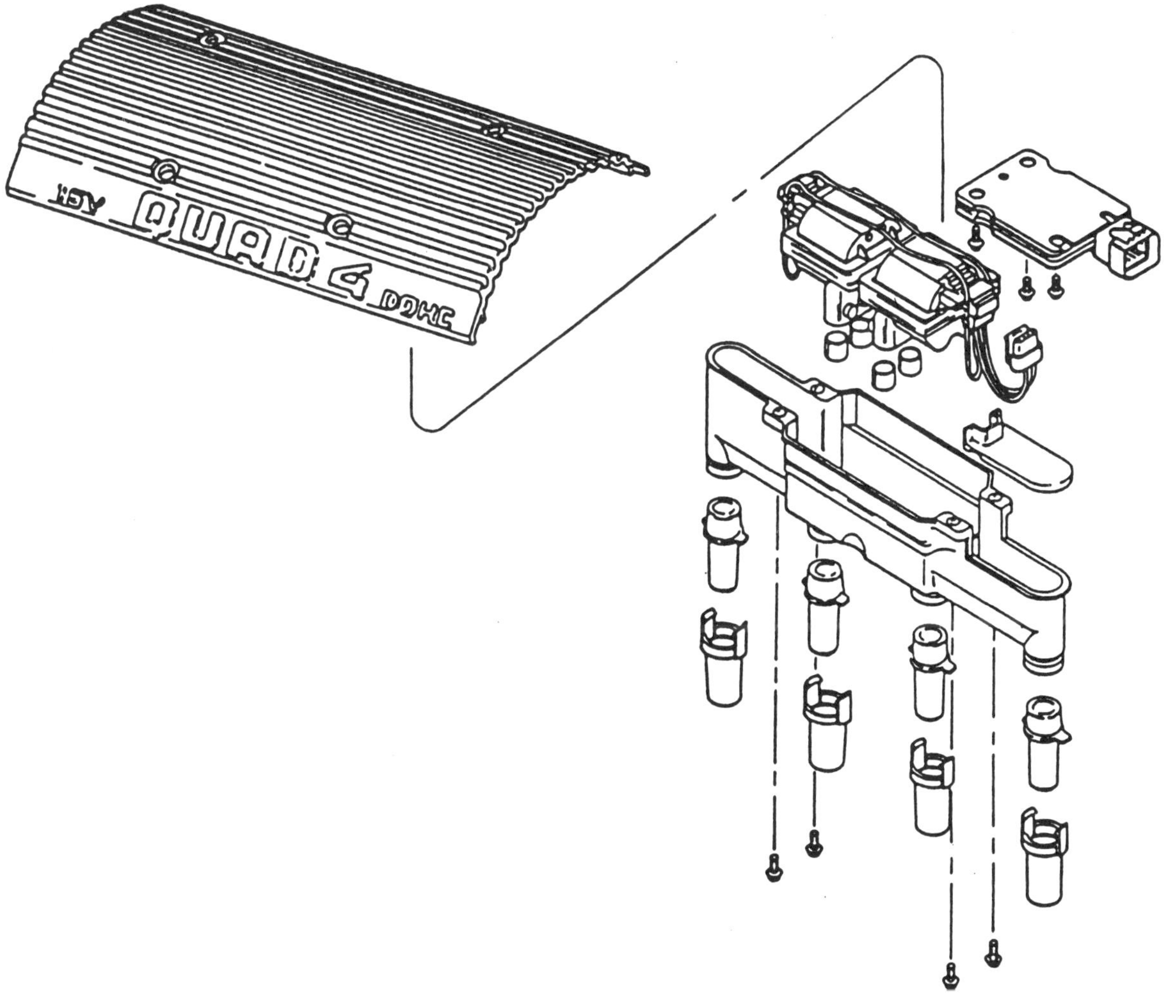

Courtesy of AC-Delco

GM Ignition System Evolution

Cont.

In summary, the GM ignition systems have evolved throughout the 70's and 80's to a sophisticated array of high energy distributor type and distributorless ignition systems. Both types of systems are interfaced with the ECM for timing and retard control through the EST and ESC subsystems. Distributorless ignition systems seem to be the the wave of the future. There will, however, still be engines produced with distributors for some time. The main difference between the two types of ignition systems is the fact that the distributor type systems are single coil fired and the distributorless systems are multiple coil fired systems. The testing philosophies involved for testing either type parallel each other significantly. Many of the test equipment manufacturers have developed special adaptors for the existing engine testers on the market. Some of these adaptors are designed to interface with the ignition primary circuits and the related ECM control circuits and some are designed to interface with the ignition secondary only. Choose your tools and testers carefully in order to prevent the mistake of choosing special testers which have limited application. The test procedures used in the following pages generally require only basic test tools. Choose other methods if you like, but realize that the system functions can be verified without expensive testers in almost all diagnostic situations.

All of these systems are some form or another of high energy ignition. The secondary voltage generated is very high and caution should be used when testing the system. The initial triggering signal, whether generated as an AC voltage or as a DC pulse is absolutely mandatory for not only ignition system operation, but also for the ECM injector pulses. In order to prevent confusion, the following test procedure pages have been separated into two main groups. The first group deals with testing the distributor type systems. The second group deals with testing the distributorless type systems.

Testing HEI Distributor-Type Systems

The technician will encounter troubleshooting situations on GM vehicles with HEI distributors which can basically be classified as one of two problems. These are:

1. No-Start
2. Ignition-related performance

The following procedures are recommended sequences for diagnosis when either of these two situations arise.

NO-START

1. Test the system for power supplied (Key "ON" and "CRANK")
2. Test the trigger signal (AC or DC Signal)
3. Verify whether or not the coil will fire (Open/Shorted Windings)
4. Inspect for coil firing to ground (Arc Marks or Burns)
5. Verify the module switching ability (Tach Signal)

To test for power supplied to the system, hook up a voltmeter to the B+ terminal of the coil. On the unitized distributors, this terminal will be on the distributor cap protrusion. On the remote mounted coils, this terminal will be on the coil itself. With the key in the "ON" or "CRANK" position, the voltage should be at or near full battery voltage.

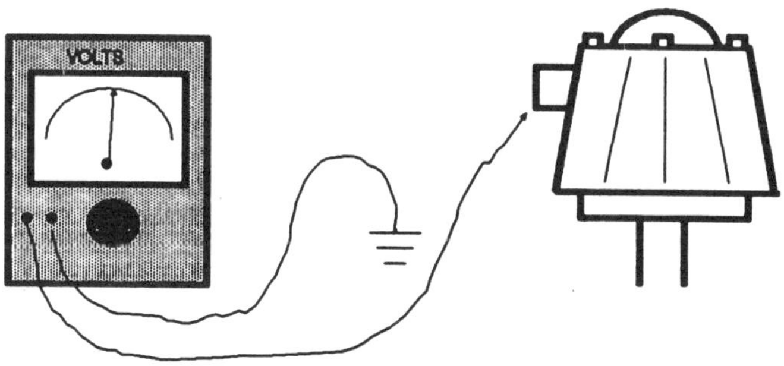

Testing HEI Distributor-Type Systems

Cont.

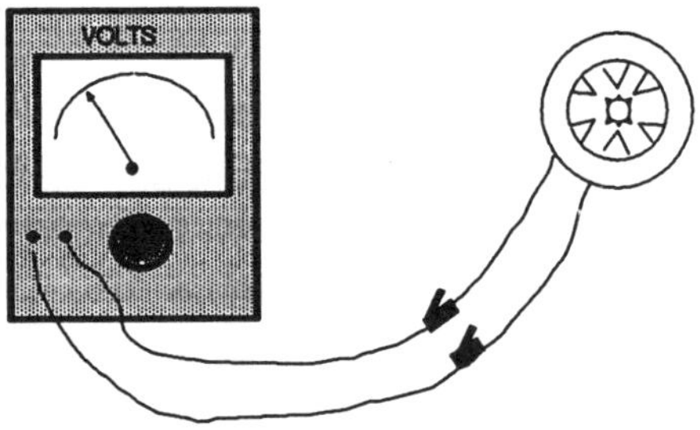

Test the coil pickup voltage with an AC voltmeter. The minimum voltage should be 0.25V AC. Usually, it will be around 1.5V to 2.0V AC during engine crank.

If the trigger signal is tested with a scope, the signal signature will look something like this.

If the distributor is equipped with a Hall Switch, the coil pickup still has to generate the AC sine wave to trigger the module and the Hall Switch has to generate the DC signal for the ECM to trigger the injectors. The Hall Switch receives power from the same B+ terminal which supplies coil B+ and module B+. The signal is being switched on the purple reference line which is attached to the Hall Switch. When the engine is cranked, test for signal line switching on the reference line with either a voltmeter or a logic probe. As the metal vanes pass through the Hall Switch, the signal should alternate from battery volts to ground or zero volts.

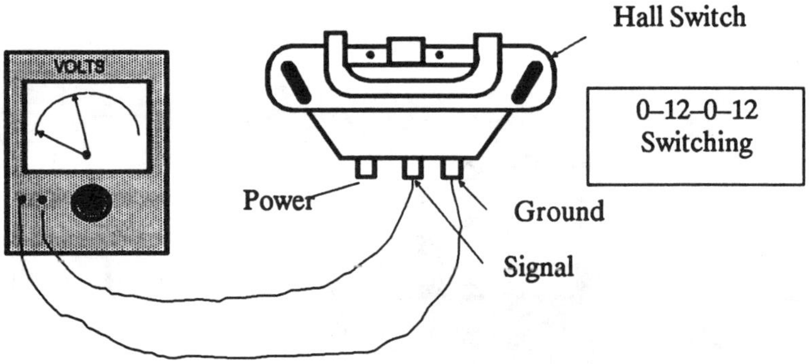

If the signal is observed on a scope, the signal signature will be a square wave which looks like this.

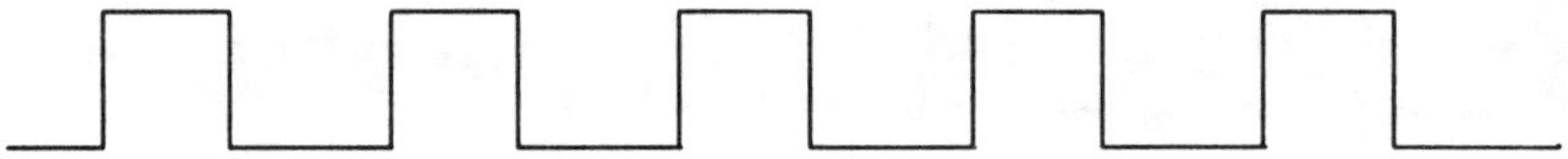

Testing HEI Distributor-Type Systems

Cont.

To verify whether or not the coil will fire, use an ohmmeter to test the coil primary and secondary windings for continuity and resistance. The primary winding will have approximately 0.5 to 1.0 ohms resistance if it is not open or shorted. The secondary winding will have 7k to 25k ohms resistance. The design changes on GM HEI coils require that the coil secondary resistance be tested from the coil-terminal to the coil secondary carbon button and from the middle ground terminal to the coil secondary carbon button. If both readings are infinite, then the coil secondary is open. If both the primary winding and the secondary windings have continuity and are within specs, then the coil is not open or shorted and is not the cause of the NO-START. Be careful to observe the cap and rotor for carbon tracks or signs that the secondary voltage is arcing through to ground.

If both readings are infinite, then the coil is bad

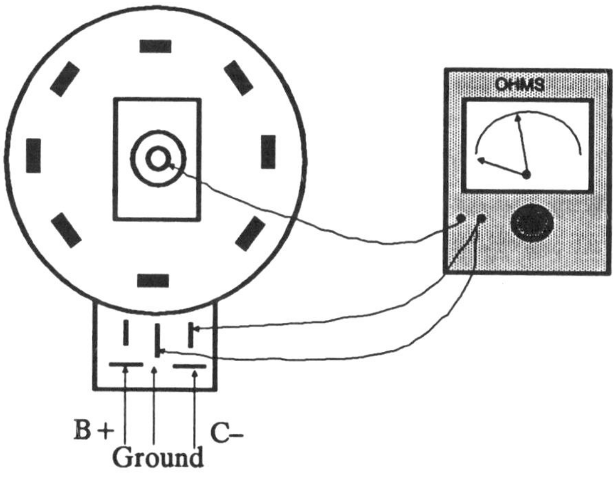

If the power supplied is OK, the pickup is generating adequate AC trigger voltage, and the coil is not open or shorted, then we can conclude that the module is causing the problem by not switching or triggering the primary circuit. Therefore, when we hook up the primary tachometer to the coil negative terminal and crank the engine, if the RPM signal is generated, then the module, pickup coil, and the primary winding of the coil are OK. We can also conclude that if the primary circuit will generate an RPM signal, then the NO-START problem is secondary-related as far as the ignition system function is concerned.

The one note of caution is to be careful with HEI/EST. If the ECM puts 5.0V on the bypass line or if the EST line is shorted to ground, then the ignition will not start the engine anyway. So, on HEI/EST, always disconnect the four-wire distributor connector and retest during crank, just to quickly verify that neither of those two scenarios is occurring.

Testing HEI Distributor-Type Systems

Cont.

POOR ENGINE PERFORMANCE – IGNITION RELATED

The best method of verifying secondary ignition system integrity on distributor equipped engines is with an oscilloscope. The basic principles for verification are:

1. Proper system polarity
2. Voltage available
3. Voltage required
4. Voltage required under load

Polarity verification is instant by observing the pattern to see if the firing lines are above the zero scope scale line. Voltage available is verified either from an open circuit test or by observing a timed firing line. If the open circuit test is used, the observed available voltage should exceed 35kV. If the timed firing line test is used, the burn time should be at least 1.5 milliseconds.

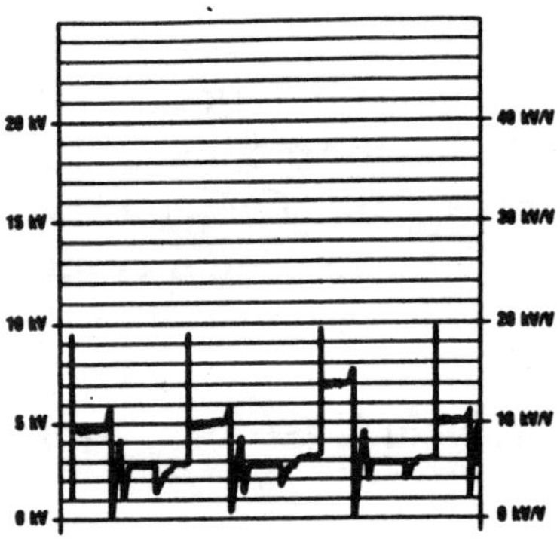

Polarity OK

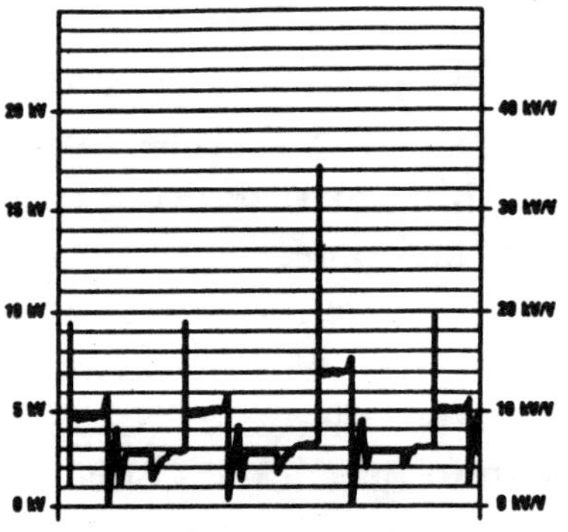

35 KV or more

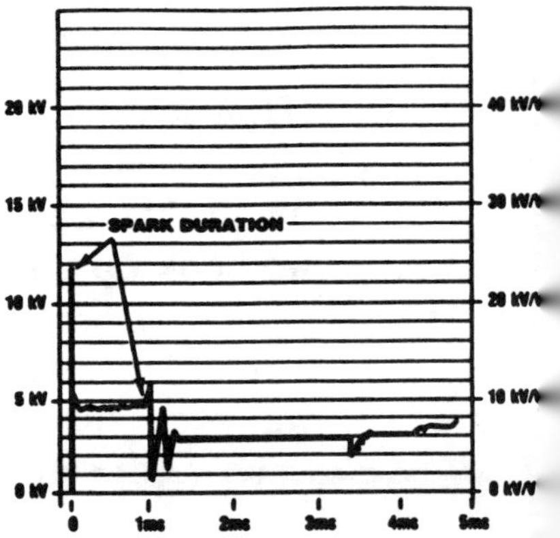

1.5ms or more

Testing HEI Distributor-Type Systems

Cont.

Load the demand on the secondary circuits by snapping the throttle plates open rapidly or, better still, load the engine with a dyno or road load unit. The firing lines should all rise at the same rate. The average firing requirements should not exceed 18kV and there should be no more than a 15% variation between any two cylinders.

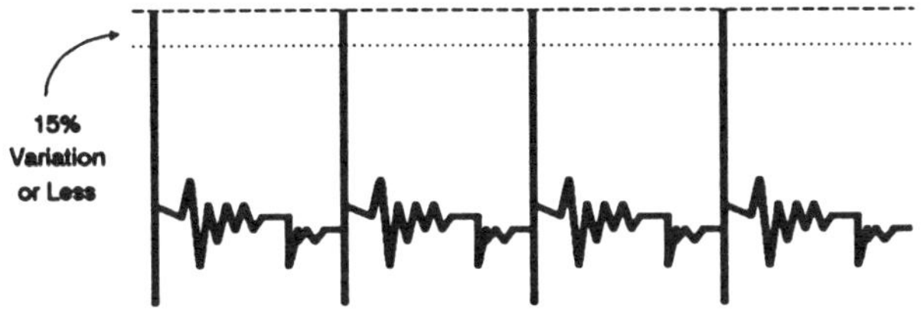

Don't forget the effect that compression has on cylinder resistance and also remember that air/fuel ratio affects the kV requirement. The other measurements you take during the basic engine systems test should adequately inform you of these factors.

Testing GM Distributorless Ignition Systems

The first distributorless ignition system used by GM was called CCCI or C^3I. It was used on the 3.8L V6 Buick engine. The first version was called Type I. This label referred to the type of coil pack used. Three coils were integrated into one assembly. If any of the coils failed, the whole assembly had to be replaced. The Type II C^3I featured three separate coils, each individually replaceable. The module assembly is serviced as a separate unit and has the coil pack bolted to it. The unique difference between the Buick C^3I system as compared to the DIS system used on the Chevrolet engines is the fact that the C^3I system used Hall effect sensors which have had a history of service failure in this application as well as other industry applications for Hall effect sensors. These Hall effect sensors must be supplied power in order to operate and the output signal will be on-off voltage pulses averaging between 6 and 8 volts.

The 3.0L V6 Buick engine uses a module and coil pack very similar to the 3.8L. The difference is the combinaton cam/crank sensor which has two slots interrupted by the metal tabs on the crankshaft pulley. The output signals are approximately 4 to 6 volts.

The 2.0L L4, the 2.2L L4, the 2.5L L4 and the 2.8L V6 engines use DIS, which stands for Direct Ignition System. On the 2.5L engine, the magnetic crank/cam sensor is inserted in a hole through the block below the coil/module assembly. The 2.0L L4, the 2.8L V6 and the 3.1L V6 have their sensors remote mounted. These magnetic sensors are similar to the coil pickup used in previous GM distributors. The signal strength can be measured with an AC voltmeter. The actual signal timing and separate synch signals can only be seen using a low voltage scope such as a lab O-scope.

The technician needs to understand where the triggering signal is generated. Then the decision to test these signals will be directly affected based on test access availability. Remember, the ECM is a digital computer and cannot process an analog signal. The module contains an A/D converter to process the AC signal into a DC signal

Testing of the secondary coils and wires will closely resemble previous secondary ignition system testing. However, proper precautions must be observed to prevent system damage or misdiagnosis. Some diagnostic analyzers can be partially modified to make certain tests, but good judgement must be used in order to evaluate cost-effectiveness.

Testing GM Distributorless Ignition Systems

Cont.

Computer Controlled Coil Ignition - CCCI – 3.0L V6 and 3.8L V6

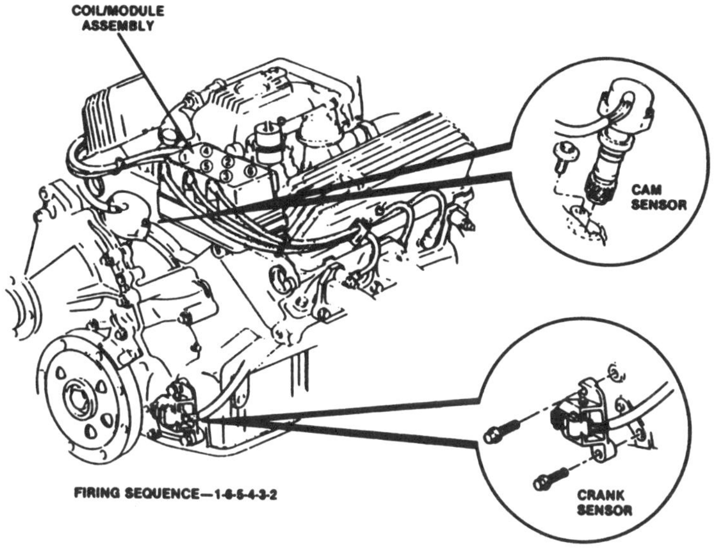

Courtesy of AC-Delco

The early 3.8 Turbo used the crank/cam sensor design and placement as shown above. The subsequent 3.8L V6's use a crank sensor mounted near the balancer and a cam sensor mounted in the timing cover as shown below. The 3.0L used a combination sensor know as the crank/cam or crank/synch sensor. It is located at the balancer. The 3800 V6 uses a dual crank sensor which generates a 3x and an 18x crank signal.

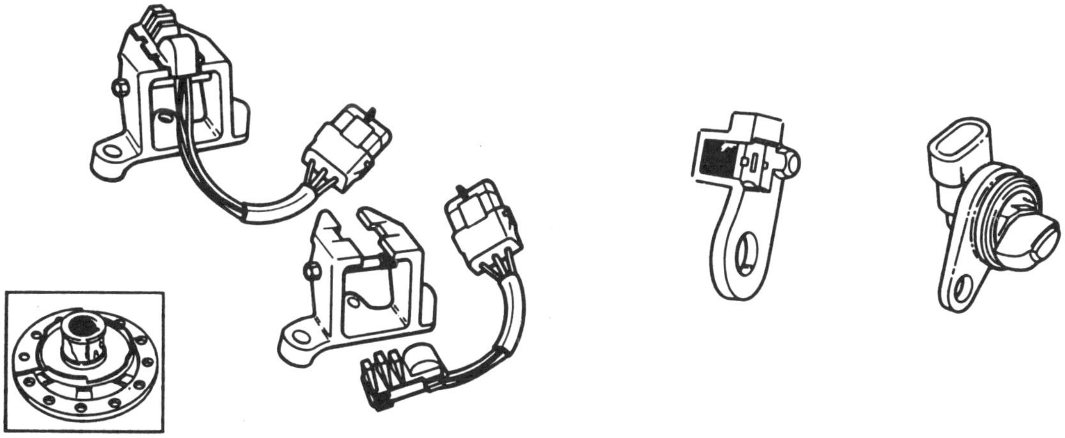

Courtesy of AC-Delco

Both sensors are Hall effect devices, supplied power by the module, which produce a DC signal strength average of 6-8 volts when the engine is running. The sensors must be powered by voltage from the module. The Hall switches cause the signal line voltage to "toggle" from high to low to high as the metal vanes interrupt the field.

Testing GM Distributorless Ignition Systems

Cont.

Computer Controlled Coil Ignition – CCCI

All of the CCCI systems will use either a Type I or Type II coil pack as seen below.

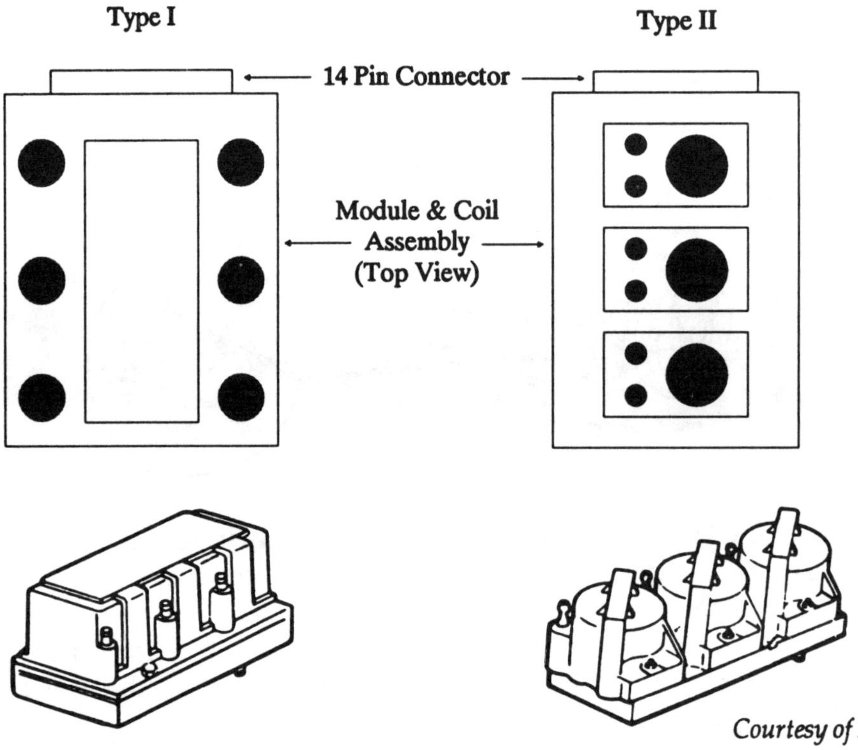

Courtesy of AC-Delco

The modules used with Type I and Type II each have 14 pins, yet they are not interchangeable.

Remember, the 3.0L and the 3.3L engines use a combination sensor called a crank/cam or crank/synch sensor. This is still a Hall effect device and will also produce a DC signal or 6 to 8 volts with the engine running.

For a no start diagnosis, is the system receiving battery volts? Are the crank sensor and cam sensor "toggling"? If not, are they receiving power from the module? Are the coils open or shorted? The coil resistance across the primary should be .5-1.0 ohms. The resistance across each secondary winding should be 10k-13k for the type I coil pack and 5k-7k for the type II coils. This resistance should be measured when the coils are not cold. Watch for signs of arcing to ground or arcing between primary and secondary.

Testing GM Distributorless Ignition Systems

Cont.

Computer Controlled Coil Ignition - CCCI

The 3.8L V6 is now sold as a 3800 V6 engine. The CCCI system used on the 3800 is called Type 3 Fast Start. The crankshaft sensor is a dual sensor which generates a normal 3x crank signal and an 18x crank signal which allows the module to set up the coil firing sequence quickly. This 18x crank signal can also be used by the ECM and module as a substitute cam signal in the event of a cam signal failure. Previous 3.8L applications would not start if the cam sensor failed. The illustration below shows the typical dual Hall Effect sensor being interrupted by a dual set of metal vanes on the crank pulley or balancer.

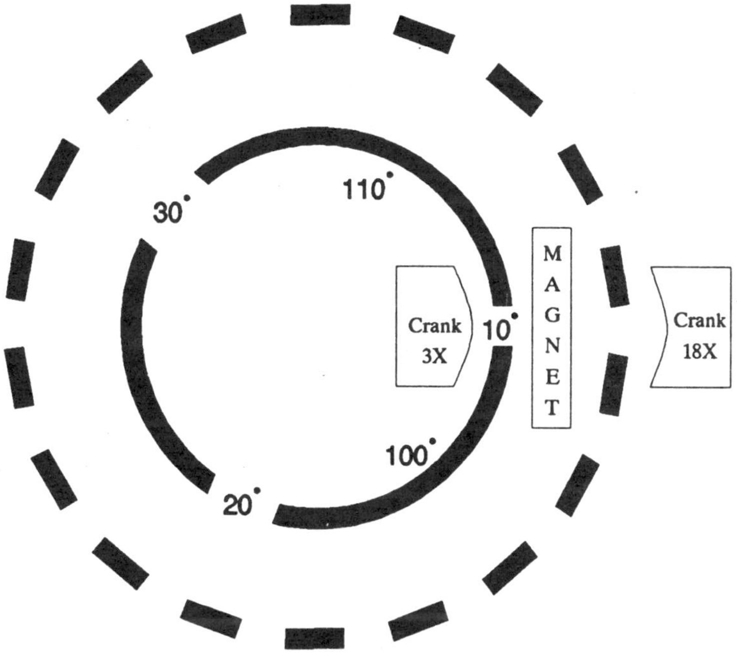

The typical signal signatures which the CCCI Hall Effect sensors can generate are illustrated below. The 3x signature will be used on all CCCI. The 18x signal signature will appear only on 3800's. The CAM signal signature will be used on 3.8's and 3800's. The synch signal generated on the 3.0L and the 3300 is the same as the CAM signal signature.

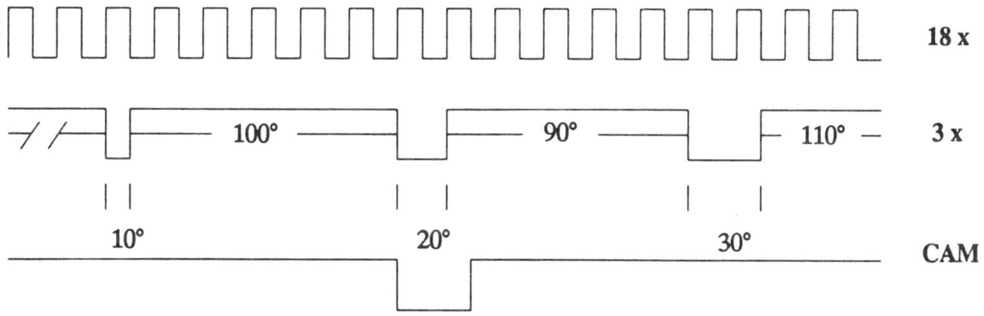

Testing DIS (Cont.)

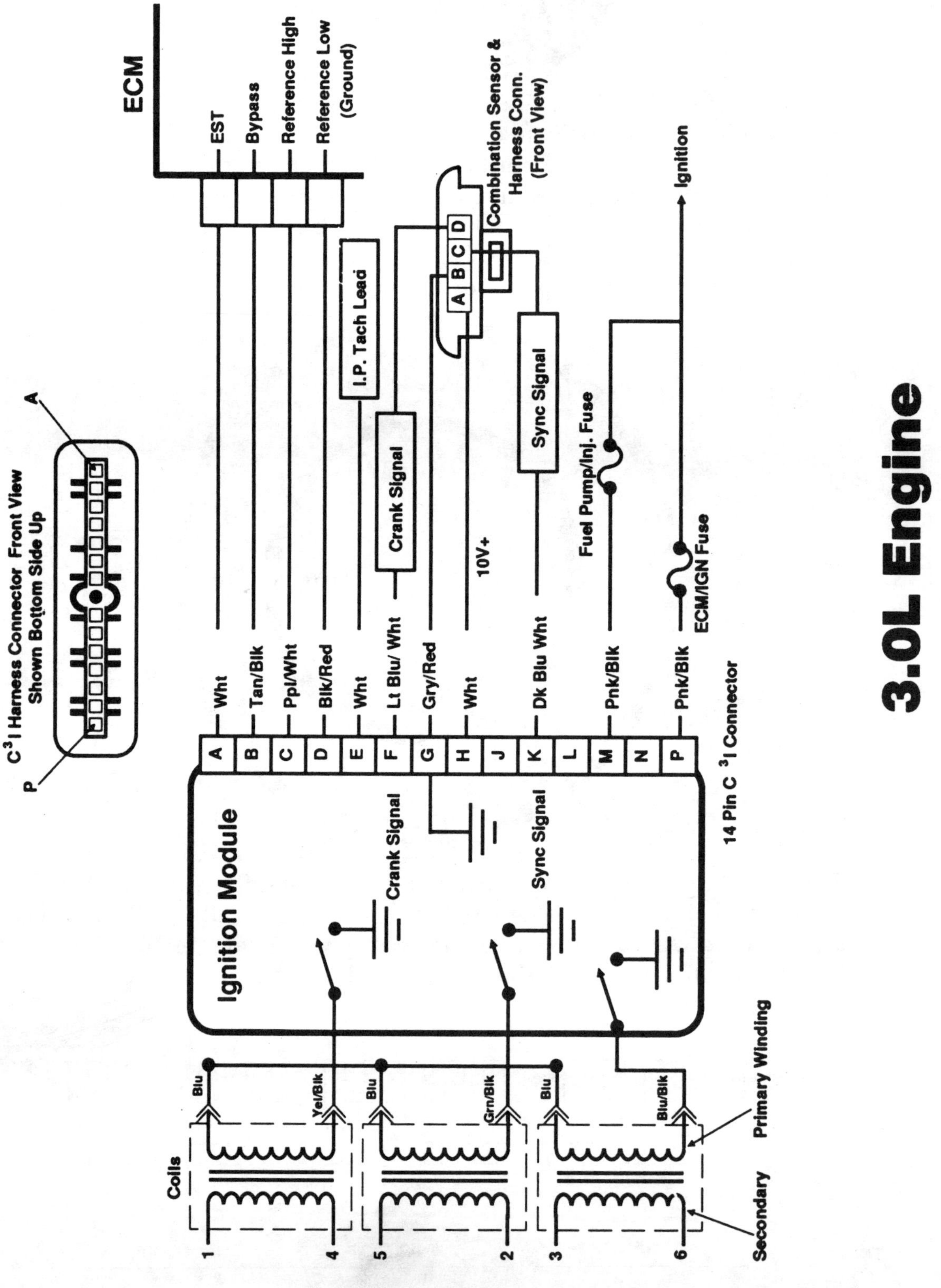

Testing DIS (Cont.)

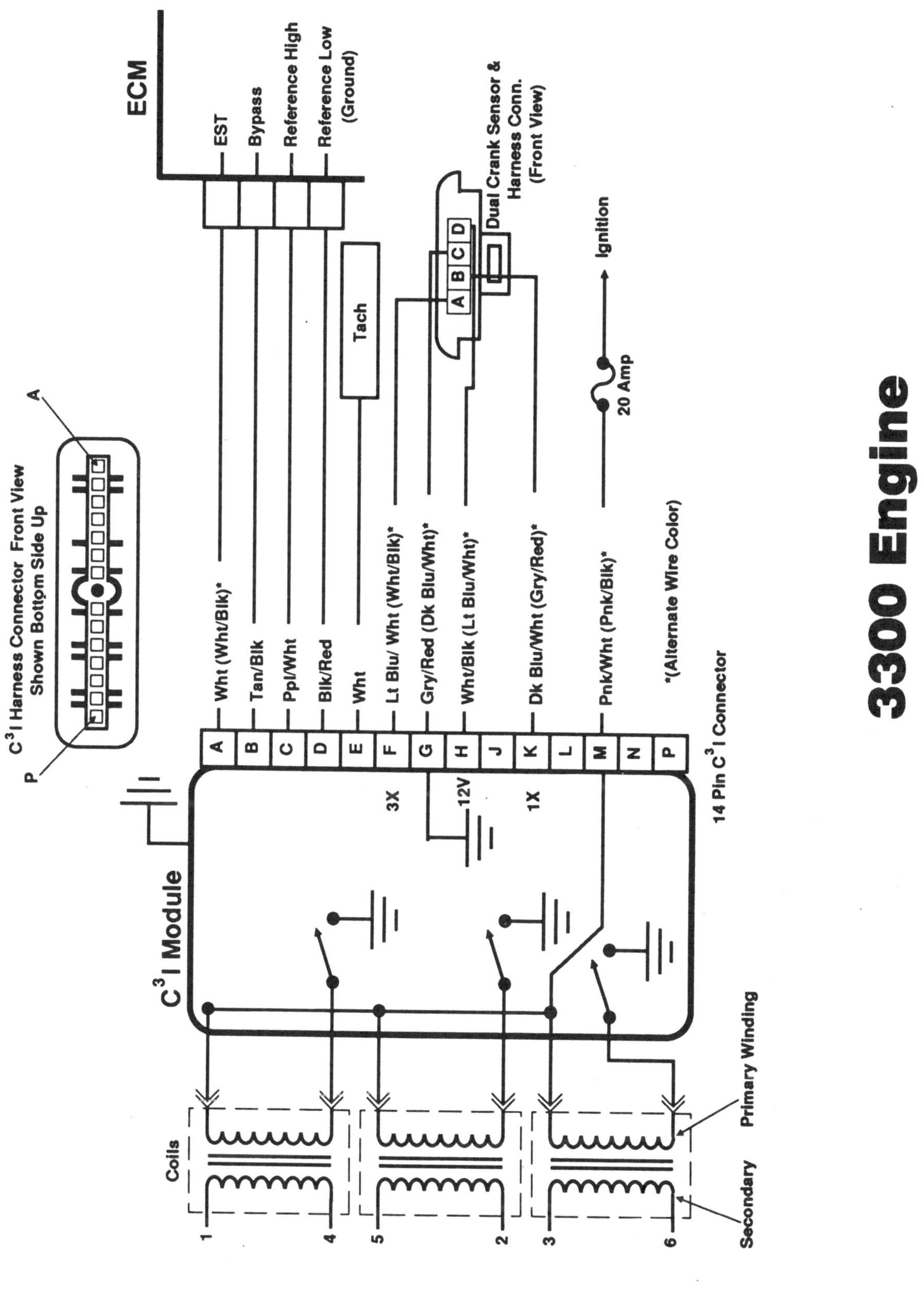

Testing DIS (Cont.)

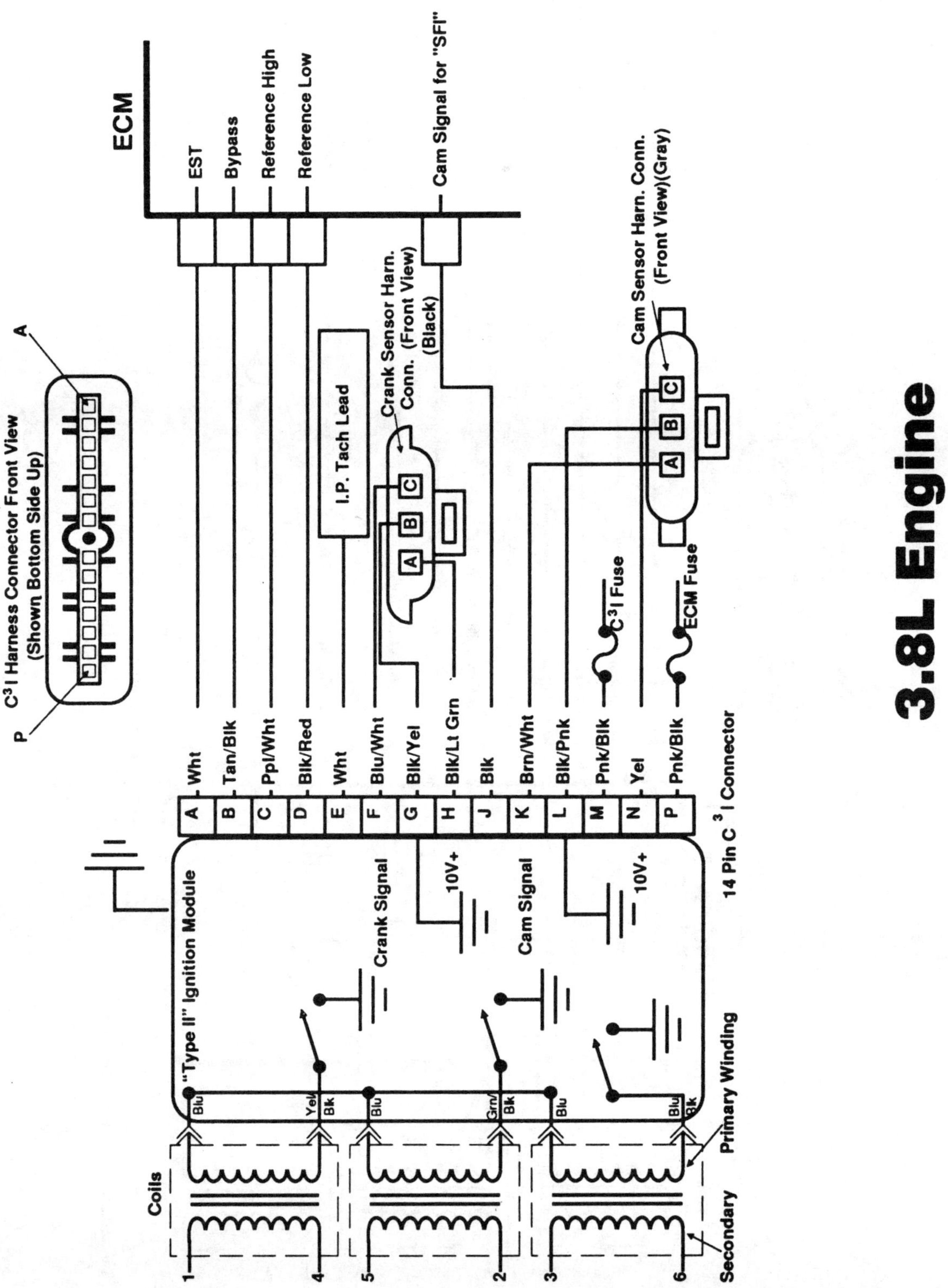

Testing DIS (Cont.)

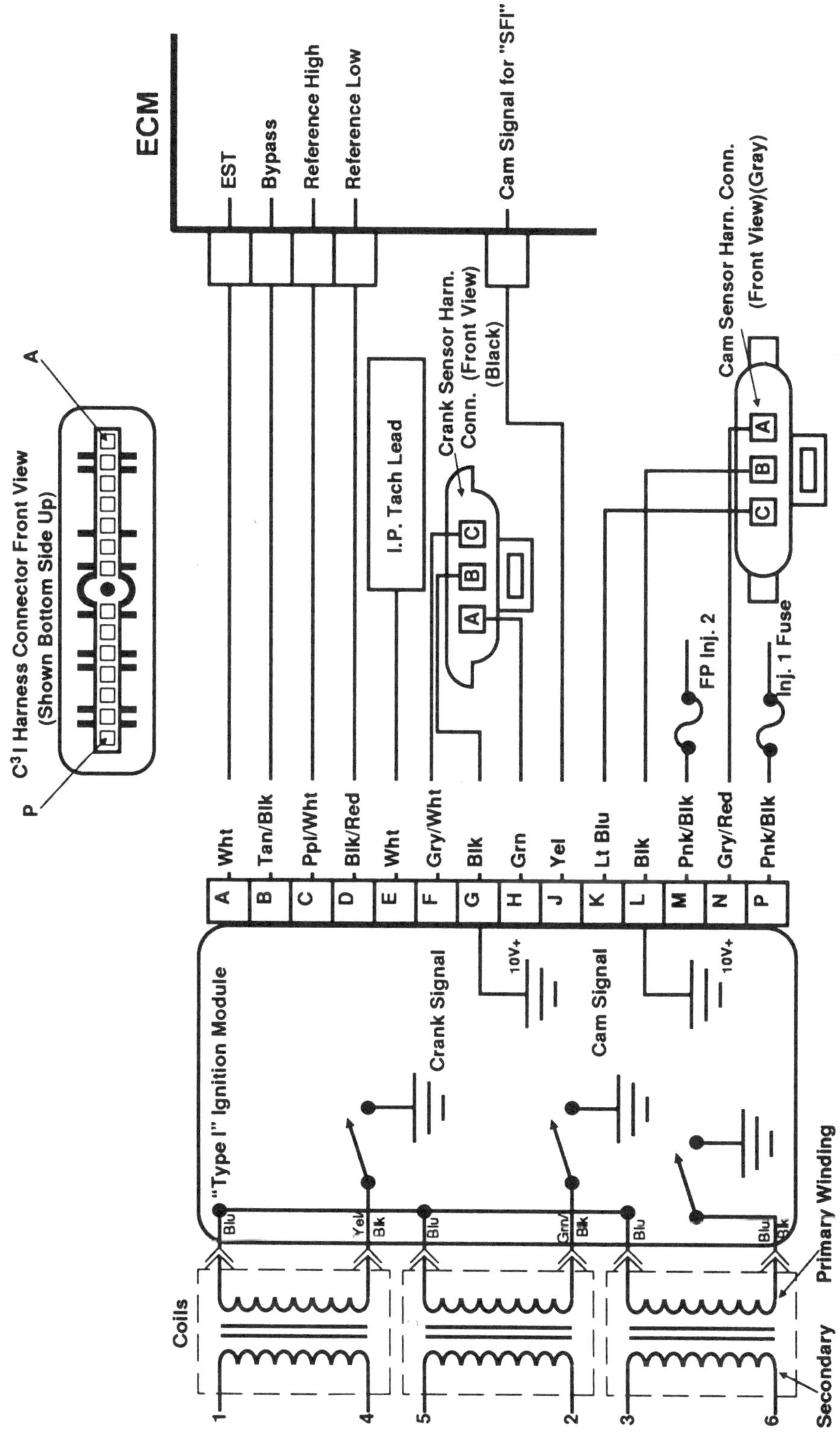

3.8L Turbo

Testing DIS (Cont.)

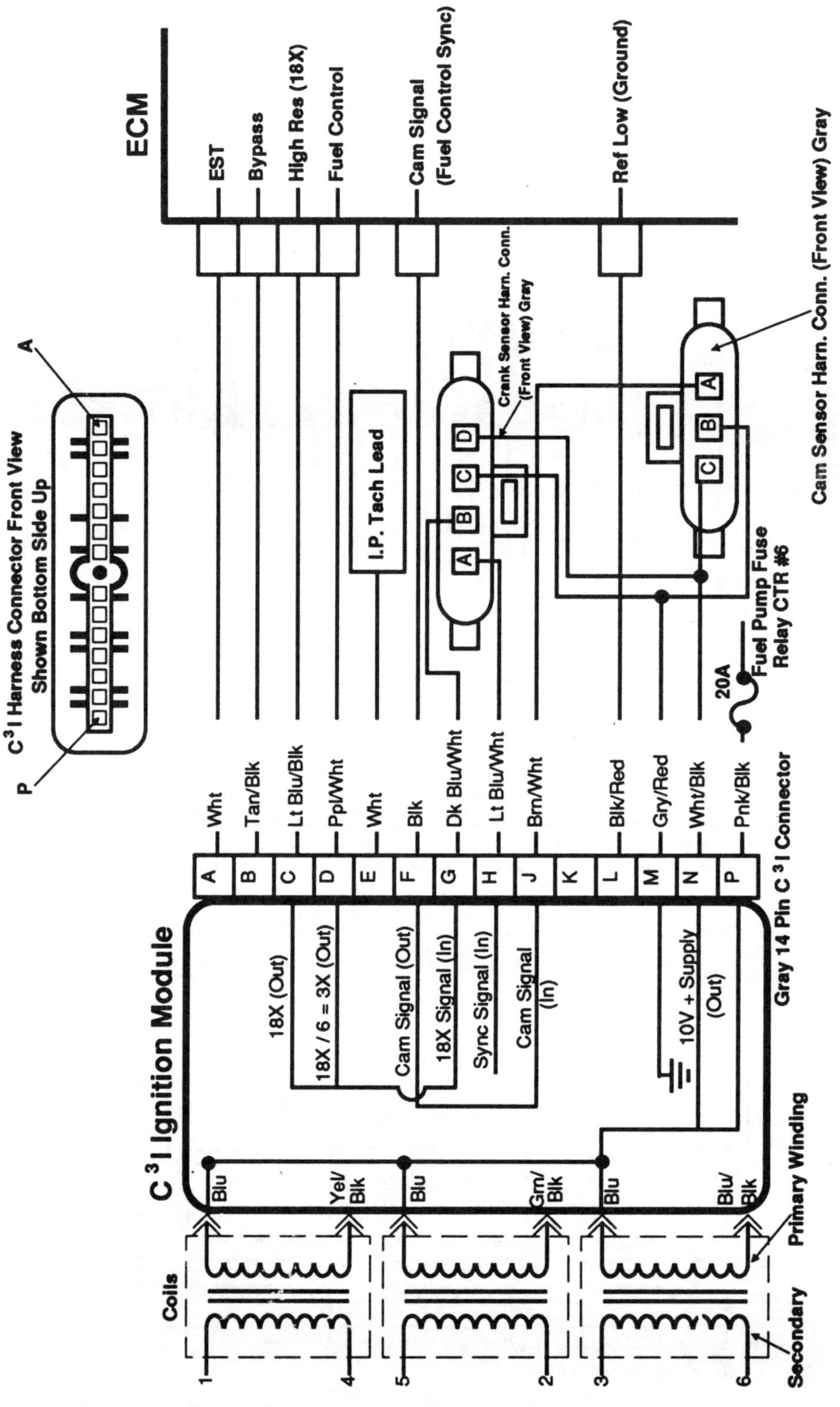

3800 Engine

Testing GM Distributorless Ignition Systems

Cont.

Direct Ignition Systems

The 2.0L, 2.2L, 2.5L, 2.8L and 3.1L Chevrolet/Pontiac engines use a version of distributorless ignition systems called DIS which stands for Direct Ignition Systems. The trigger signal input on these systems are magnetic coil pick-up style voltage generators mounted through an access hole in the engine block. The signal is generated when the reluctor wheel on the crankshaft passes its notches across the pick-up sensor. There is no adjustment for the air gap on this sensor. The sensor is remote mounted on all the engines except the 2.5L. On the 2.5L, the sensor is mounted directly below the DIS module. The signal generated is an AC sinusoidal wave form of approximately .5V AC to 100.0V AC depending on the engine RPM. The signal strength can be measured by an AC voltmeter or a DC voltmeter with a diode inserted in one of the test leads. The signal clarity of cleanness can only be verified by observing the wave form on an O-scope.

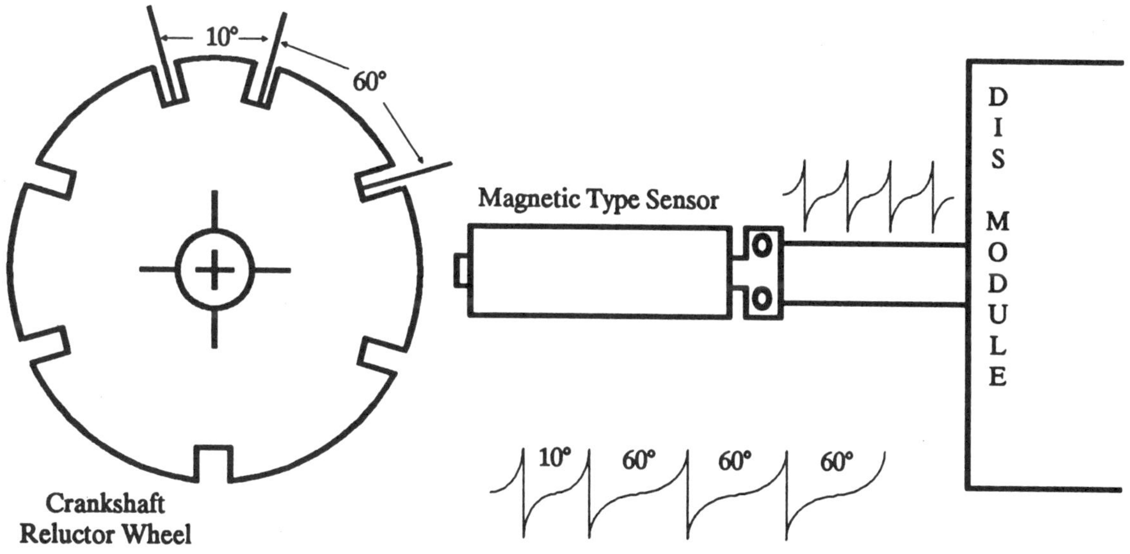

Testing GM Distributorless Ignition Systems

Cont.

Direct Ignition System

Since both the ignition module and the ECM are digital devices, the AC crank/synch signal is unusable in its initial form. The DIS module contains A/D converter circuits which convert the AC inputted signal to a digital signal. This signal is used by the module to fire the primary circuit transistors when the system is functioning in the module mode. This converted signal is also sent to the ECM as "reference" so the ECM can use it to fire the fuel injector(s). The ECM also uses this digital Reference signal to fire the DIS module transistors when the ECM is controlling the timing of the engine. This time-shafted "reference" signal is issued by the ECM and sent to the DIS module as EST. To test the "reference" line of the "EST" line, the technician can use a logic probe to verify that the digital pulse is present. To verify the cleanness of the signal, the signal has to be observed on an O-scope.

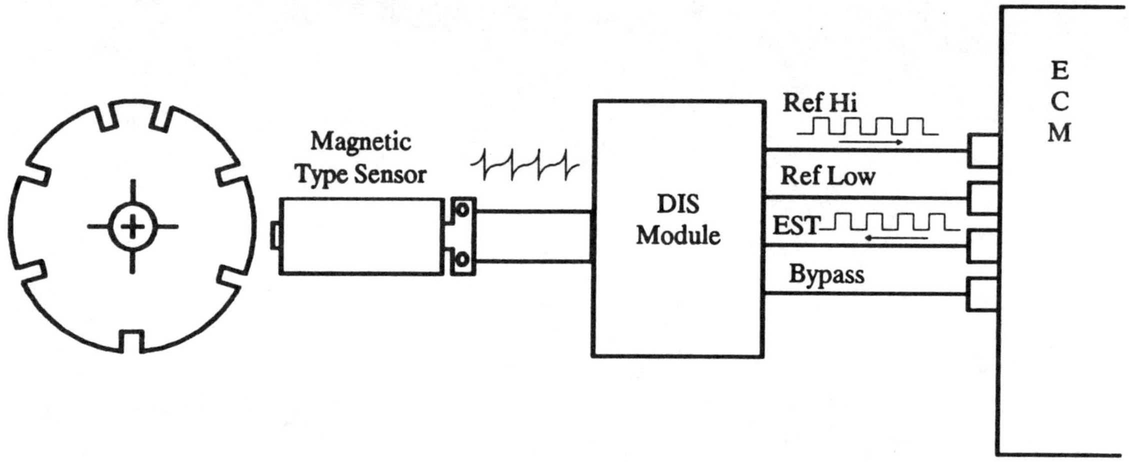

Two Modes — <u>Module Mode</u>

Bypass Low
EST Pulled

<u>EST Mode</u>

Bypass High
EST Timing Control

Testing GM Distributorless Ignition Systems

Cont.

Direct Ignition System – 2.0L

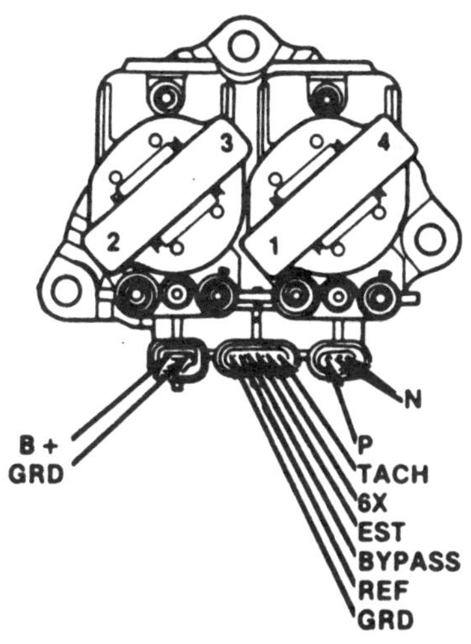

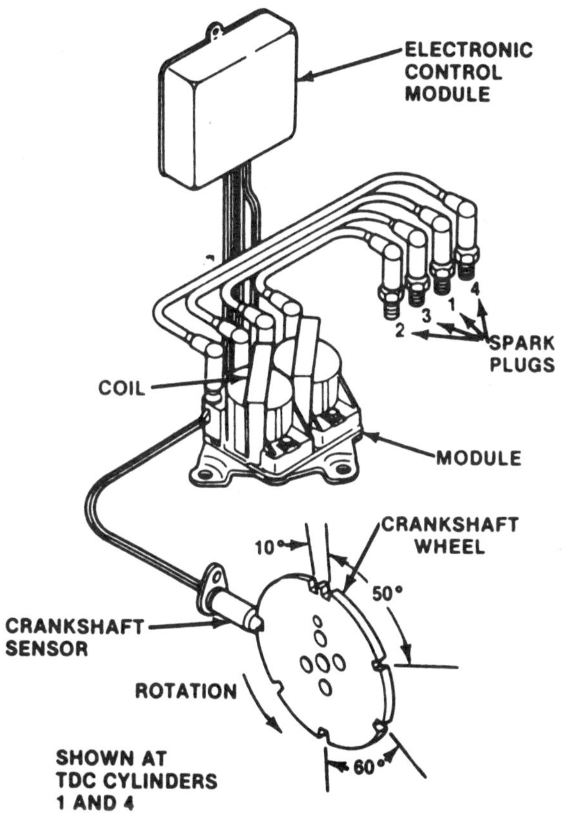

Courtesy of AC-Delco

The 2.0L and 2.2L L4 Direct Ignition Systems use a magnetic generator type crankshaft sensor. It generates its own AC voltage which is delivered to the module via the P and N terminals. This signal increases in intensity and frequency as engine RPM increases.

Testing GM Distributorless Ignition Systems

Cont.

Direct Ignition System – 2.5L

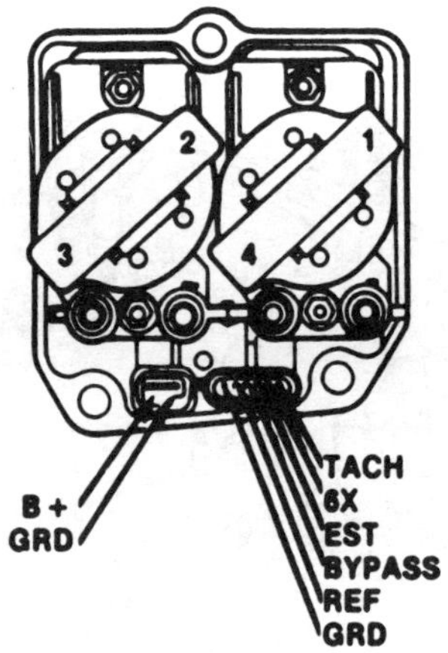

Courtesy of AC-Delco

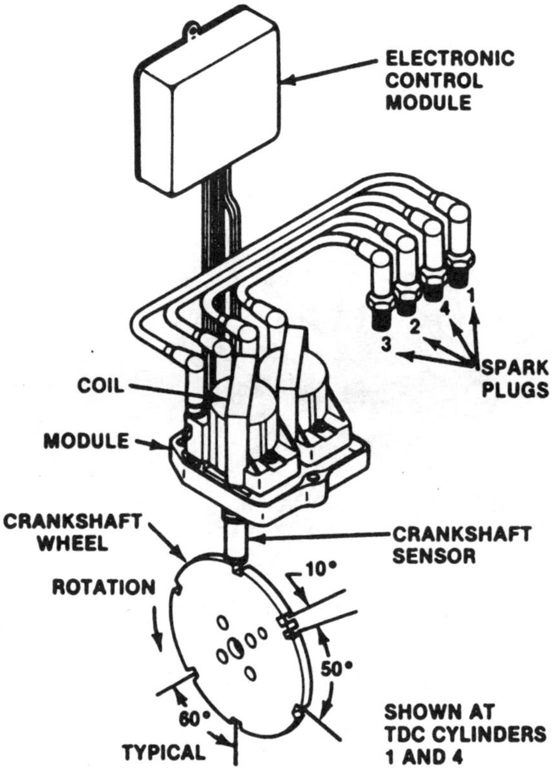

The 2.5L L4 Direct Ignition System also uses a magnetic generator type crankshaft sensor. The only difference is that the sensor is located directly under the module/coil pack. Therefore, the P & N terminals are on the bottom side of the module which has to be removed in order to test crank sensor.

Testing DIS (Cont.)

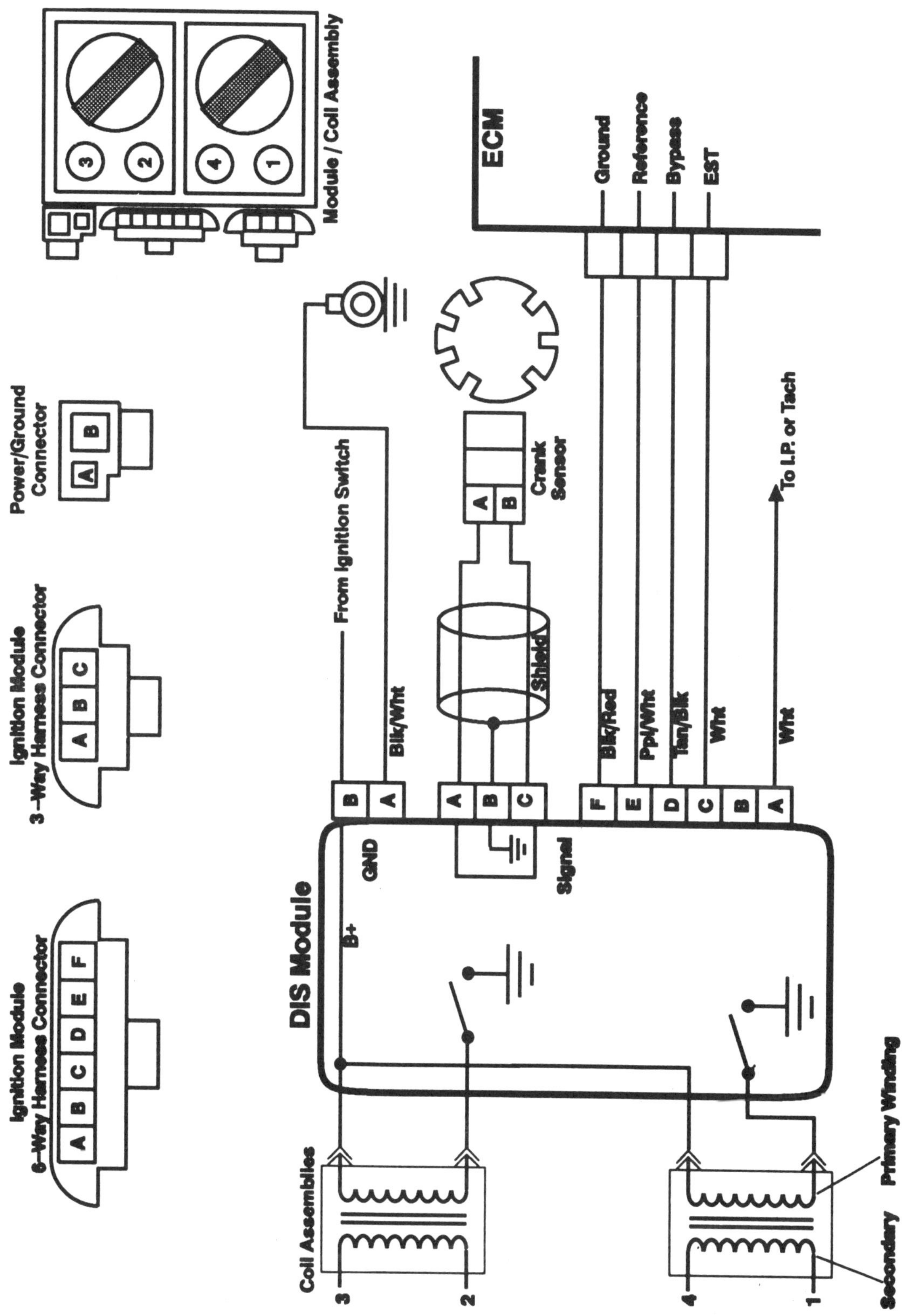

2.0L & 2.2L Engine

Testing DIS (Cont.)

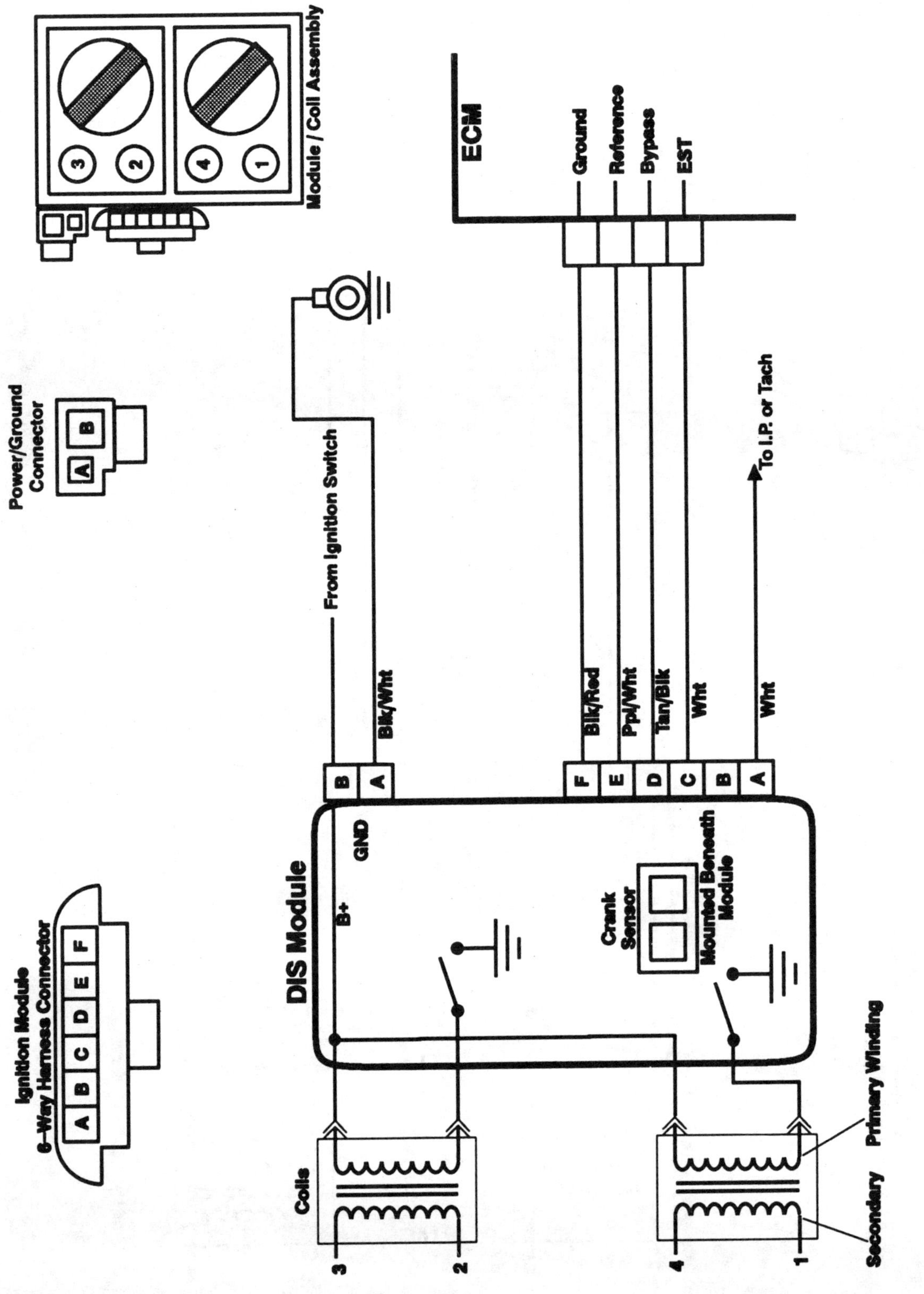

2.5L Engine

Testing GM Distributorless Ignition Systems

Cont.

Direct Ignition System – 2.8L and 3.1L

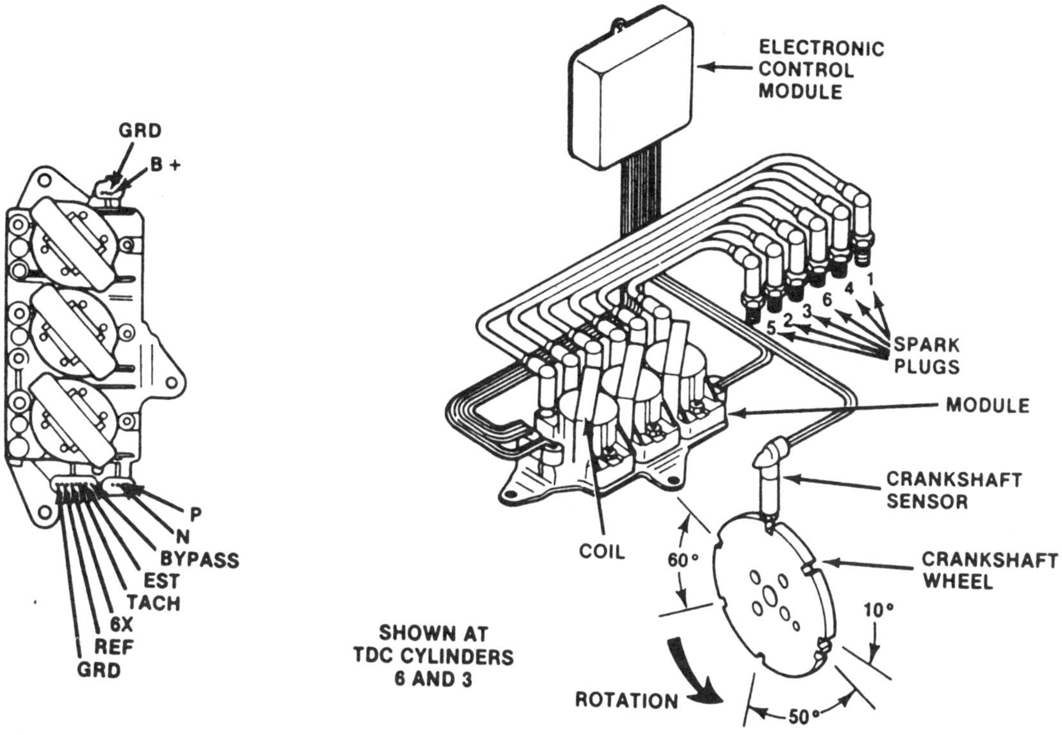

Courtesy of AC-Delco

The 2.8L V6 Direct Ignition System also uses a magnetic generator type of crankshaft sensor. Notice that the inputs and outputs of each DIS module are the same. The only variation is their physical location on the module.

Testing DIS (Cont.)

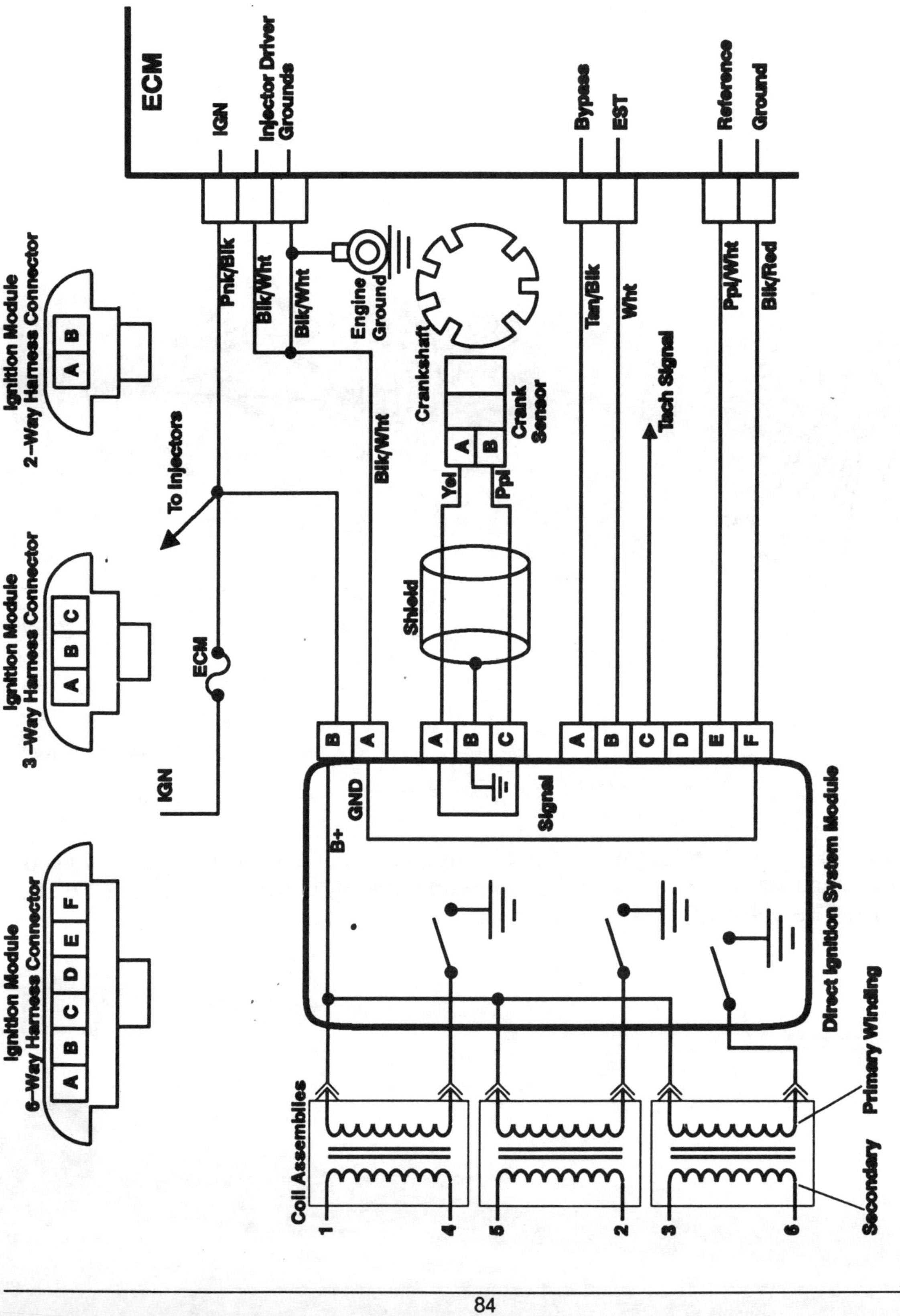

2.8L & 3.1L Engines

Secondary Circuit Testing

The ignition coils used on Direct Ignition Systems should have a primary circuit resistance of less than one (1) ohm. The secondary winding should test out at 5k to 7k resistance. The voltage induced into these coils is so high, they are having quite a problem arcing through the insulation. Anytime misfire is occurring, carefully inspect the coils for any visual evidence of arcing to ground or between the primary and secondary.

Because these new systems represent a drastic variation from the single-coil fired ignition systems which we have dealt with in previous years, the manufacturer has developed some different techniques for diagnosis. Observing the illustration below, the technician should realize that in the spark method of coil firing, both plugs on both ends of the coil secondary are firing simultaneously. On a four cylinder engine there are two of these secondary circuits and on a six cylinder engine, there are three such circuits.

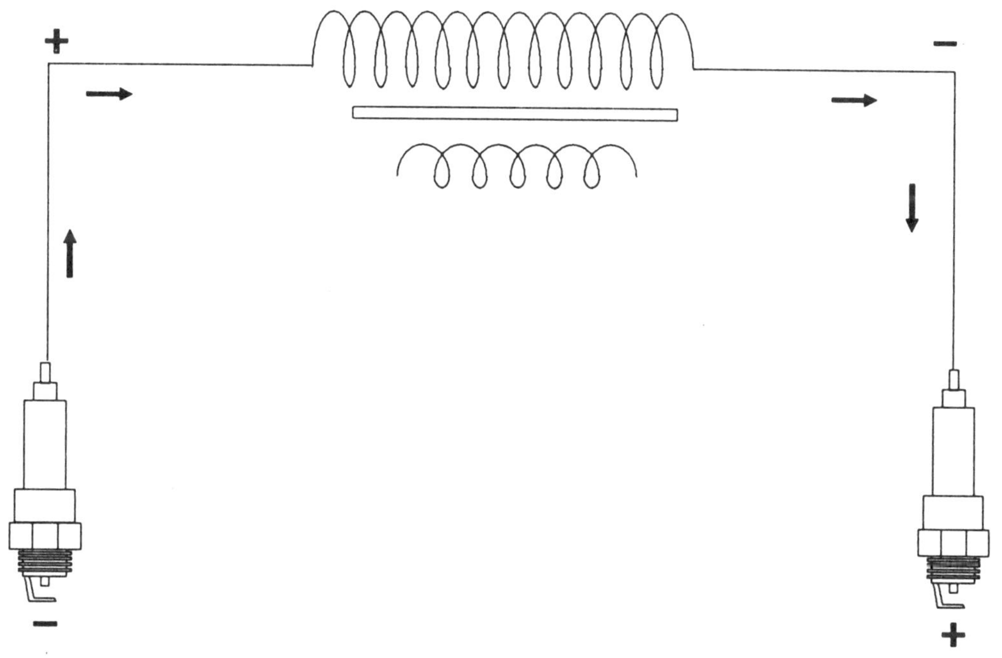

Secondary Circuit Testing

Cont.

The logical steps for the technician to follow are:

1. Verify kV available – test with HEI spark tester
2. Verify voltage required – use the grounding probe to eliminate high resistances
3. Verify voltage delivered – use timing light or firing indicator on both sides of each coil circuit.

Coil insulation integrity can be tested by placing an HEI spark tester at the end of both secondary plug wires for each coil circuit for companion cylinders. While cranking the engine, both HEI spark testers should fire simultaneously and consistently. If the spark at either tester is erratic, the coil insulation is leaking internally. Secondary voltage is seeking and finding an alternate path to ground. An ignition coil can have proper resistance readings, but if the insulation is defective, misfire will occur. An ohmmeter can not test insulation integrity.

If the spark tester on one side of the coil fails to fire completely, use a simple grounding probe to eliminate the plug, then use the plug wire to determine which part of the opposing secondary circuit has high resistance. The process of relocating coils can also be helpful in determining if the problem follows the moved coil, thereby pinpointing a coil problem. Remember, there is a primary driver transistor in the ignition module for each coil circuit.

Remember, a simple ohmmeter test can detect an open plug wire. Visually inspect for insulation leaks or deterioration. An easy power balance can be performed using short pieces of high-carbon vacuum tubing and a test light or grounding probe.

Some engine analyzer manufacturers have begun to design and release special adaptor sets for use with these new systems. But, be careful in selecting such adaptors because of high costs and relatively limited applications.

Engine Firing Orders

1.8L L4

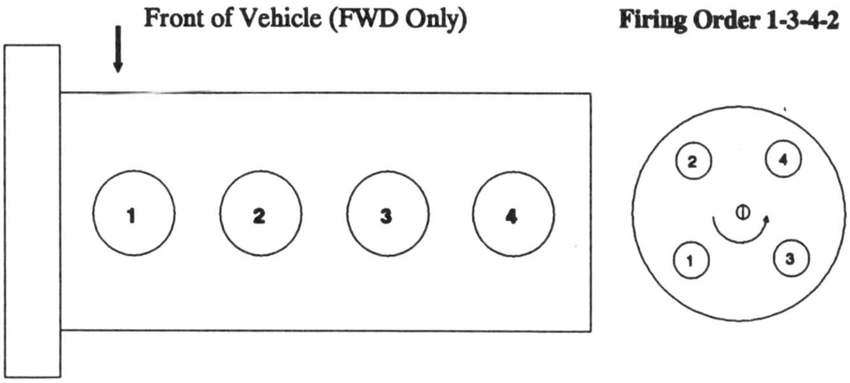

2.5L L4

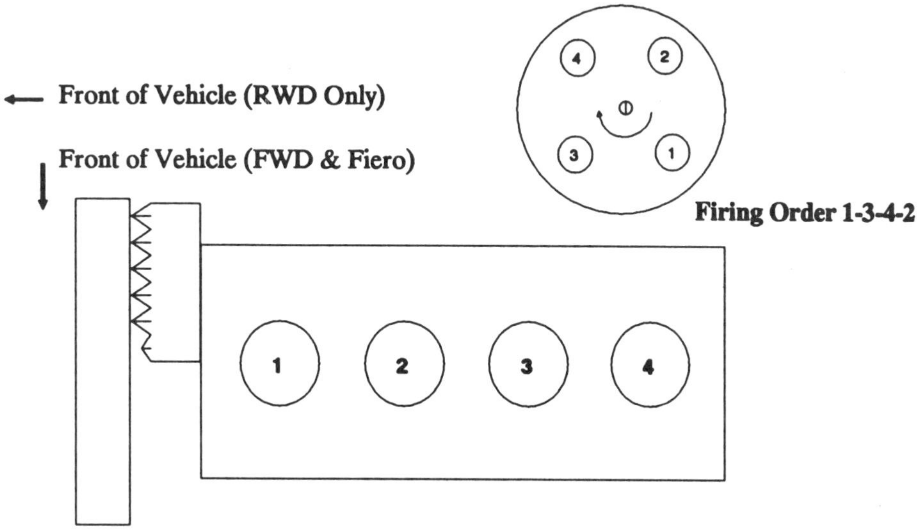

2.0L & 2.2L L4

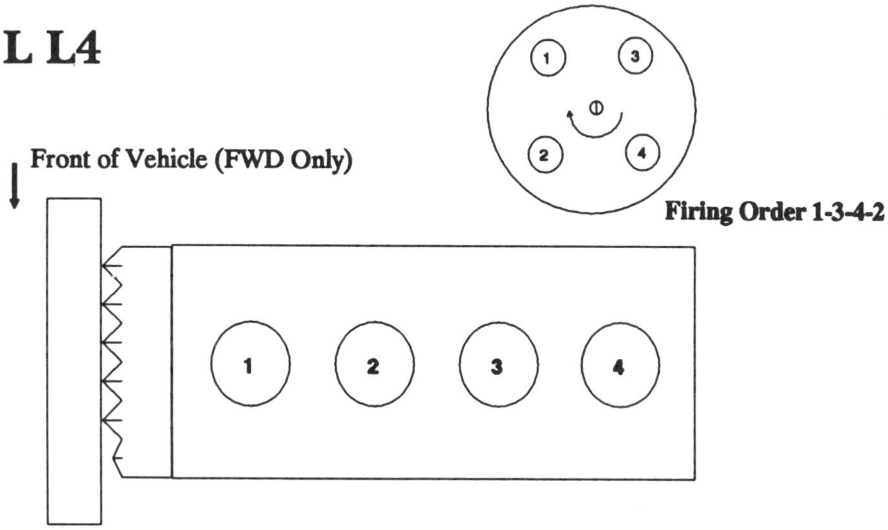

Engine Firing Orders (Cont.)

2.8L V6

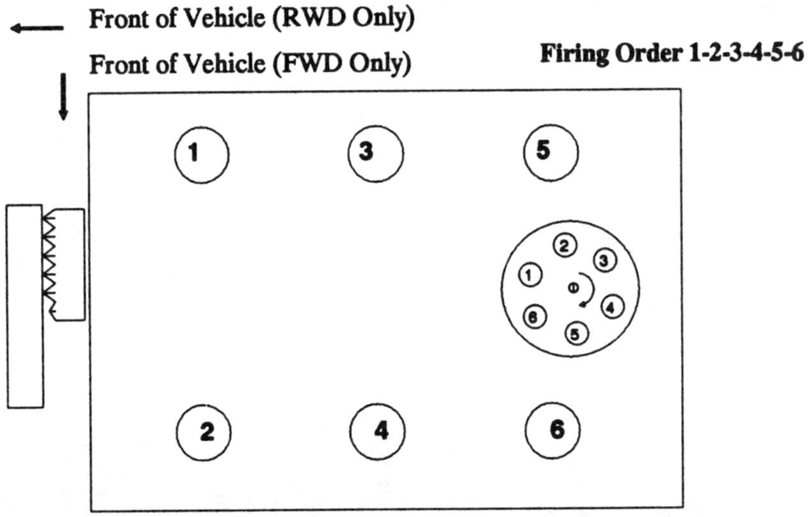

3.0L & 3.8L

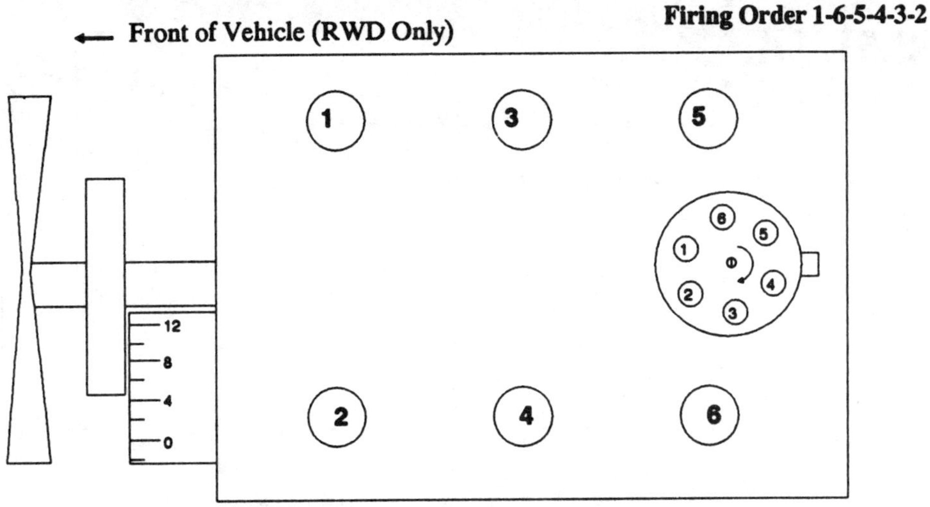

5.0L & 5.7L

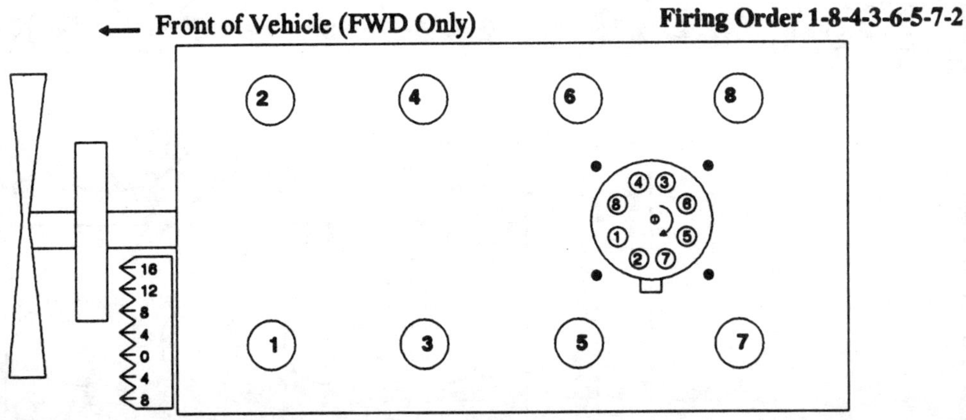

Timing Adjustment Procedures

		Ignition Timing Procedure-Step Numbers[1]					
Application	**VIN**	**1982**	**1983**	**1984**	**1985**	**1986**	**1987**
Passenger Cars							
1.8L L4 Turbo MFI	J	–	–	2	2	2	
1.8L L4 TBI	O	2	2	2	2	2	–
2.0L L4 TBI	P	–	2	2	–	–	–
2.0L L4 TBI	P	–	–	–	3	3	–
2.0L L4 H.O. TBI	1	–	–	–	–	–	–
2.0L L4 TBI	K	–	–	–	–	–	2
2.0L L4 Turbo MFI	M	–	–	–	–	–	2
2.5L L4 FWD TBI	R	2	2	2	2	2	2
2.5L L4 TBI	U	–	–	–	2	2	2
2.5L L4 RWD TBI	Z	2	2	2	2	2	2
2.8L V6 RWD MFI	S	–	–	–	3	3	3
2.8L V6 FWD MFI	W	–	–	–	3	3	–
2.8L V6 FWD MFI	9	–	–	–	3	3	2
3.0L V6 SFI	L	–	–	–	6	6	6
3.8L V6 MFI	B	–	–	–	–	–	–
3.8L V6 SFI	3	–	–	1	1	1	1
3.8L V6 Turbo SFI	7	–	–	–	–	6	6
3.8L V6 Turbo SFI	9	–	–	6	6	–	–
4.1L V8 DFI	8	4	4	4	4	4	4
4.1L V8 PFI 7	7	–	–	–	–	–	–
4.3L V6 TBI	Z	–	–	–	3	3	3
5.0L V8 TPI	F	–	–	–	3	3	3

The timing adjustment procedures for engines later than 1987 will usually be the same as for 1987 if the engine is equipped with a distributor. Most GM engines from 1988 on are distributorless, hence no adjustment is possible unless you remachine the crankshaft or make some other radical change to the hardware.

Timing Adjustment Procedures (Cont.)

APPLICATION	VIN	1982	1983	1984	1985	1986	1987
Passenger Cars cont.							
5.7L V8 TBI	8	3	3	3	—	—	—
5.7L V8 TPI	8	—	—	—	3	3	3
6.0L V8 DFI	9	7	5	5	—	—	—
Light Trucks							
2.5L L4 TBI	E	N/A	N/A	N/A	1[2]	1[2]	2
2.8L V6 TBI	R	N/A	N/A	N/A			3
4.3L V6 TBI	Z	N/A	N/A	N/A		1[2]	3
5.0L V8 TBI	H	N/A	N/A	N/A			3
7.4L V8 TBI	N	N/A	N/A	N/A			3

[1]For interpretation of step numbers, refer to Ignition Timing Procedure-Step Numbers below.

[2]California Models only.

Step Numbers

1) Disconnect 4 wire EST connector at distributor.

2) Ground diagnostic test terminal. Terminal B of ALDL.

3) Disconnect TIMING BYPASS CONNECTOR (tan/black wire). The connector may be located in one of 3 locations:
   a. near right valve cover,
   b. near blower motor, or
   c. near distributor.

4) Disconnect green reference signal connector at rear of engine, near dis tributor. Set timing with engine idling under 800 RPM.

5) Disconnect SET TIMING CONNECTOR.

6) Timing is electronically controlled and is not adjustable.

Review of Key Concepts

- To earn revenues from diagnostics, you have to develop a disciplined, orderly system of diagnostic procedures.

- Diagnostic reports are mandatory to achieve the disciplines necessary to follow logical procedures.

- Diagnostic reports are also good marketing tools.

- GM ignition systems are simple to test and easy to understand.

Table of Contents

Section III

Section III Outline

III. Troubleshooting EFI

A. After verifying the basic engine systems, then enter the self-diagnostic system.
   1. The fault codes can be obtained by counting the CHECK ENGINE lamp flashes.
   2. The codes can also be obtained by interfacing the ECM with a scan tool.
   3. Determine which codes represent HARD FAULT CODES by clearing the retrieved codes from memory and retesting the vehicle to see if the code or codes will return.
   4. Use the FIELD SERVICE MODE to determine if the ECM will go into a closed loop and verify the system's ability to manage fuel properly.
   5. While in the DIAGNOSTIC CIRCUIT CHECK mode, verify the integrity of the ECM driver circuits by testing for voltage presence and voltage change.
   6. Testing the system's ability to operate in FUEL BACK–UP control will verify whether or not the CalPack is OK.
   7. Remember that State 9, ROAD TEST STATE is better for EFI diagnosis because it does not modify ECM behavior. STATE 8, the ALDL STATE can be used if necessary.

B. After thoroughly determining which subsystems are operating and which are not, proceed with individual circuit tests.
   1. Check the ECM power leads and grounds.
   2. Check each input device and the related circuits.
   3. Check each output device and the related circuits.
   4. Determine if the ECM is inoperative by the process of logical elimination.

GM's Self-Diagnostic System

General Motors EFI engine controllers all contain a self-diagnostic system. This ability to monitor various subsystems and circuits is misconceived by many technicians. The self-diagnostic system is comprehensive in some of its abilities, yet very limited in others.

The key to utilizing this self-diagnostic capability is in understanding the diagnostic philosophy utilized by the GM engine controller designing engineers. GM basically accomplishes sensor and subsystem monitoring by the use of "continuous" self-testing. You don't have to tell or request the ECM to look for a circuit problem. Any time the ignition key is ON, the ECM is inputting the sensor signals and comparing them to pre-programmed diagnostic "windows". Some of these windows are very tight and some are very wide. The computer also utilizes the internal clock to make decisions as to when a sensor signal is out of range long enough to be considered at fault. The General Code Definitions listed in this section of the manual will illustrate the various windows and time sequences which are involved in circuit faulting.

Any time the ECM detects a window violation, the ECM is designed to turn ON the CHECK ENGINE or SERVICE ENGINE SOON lamp. This should inform the vehicle operator that the ECM has detected a fault. The ECM also has the ability to store detected fault codes in long term memory in the event that the fault does not occur continuously. The long term memory is usually cleared of the fault code if the code parameter violation goes away and does not return for a period of 50 crank start cycles.

The technician can access the fault code storage and display section of the ECM through the ALDL or Assembly Line Diagnostic Link. The ECM will flash the CHECK ENGINE lamp indicating any stored fault codes whenever the diagnostic lead is grounded. The diagnostic lead is always wired to terminal B of the ALDL connector. The other end of that wire terminates at the ECM. A better explanation of the ALDL and the diagnostic circuits follows later in this manual.

The other option for the technician concerning fault code retrieval is the use of the scanner. Since GM ECMs contain a serial data transmitter, the tester industry has developed scan tools which are designed to access that data stream. This has been both a blessing and a curse. Properly understood, this scan tool capability can be a real time saver for many diagnostic situations. It has also been a hit or miss proposition for many of the real-life circuit and sub-system problems.

GM's Self-Diagnostic System

Cont.

GM has continued to upgrade its controller diagnostic systems by adding additional capabilities each model change. The diagnostic codes pertinent to each system have steadily become more encompassing as this expansion continues. The SPECIFIC FAULT CODE DEFINITIONS are located in Section IV. These are listed according to the year model. These listings will help the technician pinpoint the actual code definition when the GENERAL FAULT CODE DEFINITIONS list more than one possible definition.

There are several notes of caution which the technician must recognize. First, identify the system carefully, especially when the scanner is going to be utilized. The ECM will not communicate properly when the system identification does not match the scanner set-up. Second, the later model systems have much better default circuit value substitutions. This means that the engine will seem to be performing quite well, when actually there may be a major problem. The third note of caution is to remember that the self-diagnostic capability of the ECM is limited and that there are plenty of things that can go wrong with GM EFI engines which will not set a code. There are also many things that can set a code which will not always mean that the engine performance is suffering significantly.

In summary, the GM self-diagnostic system is by and large a great aid to the vehicle operator and to the trained technician. The process of sophisticating the self-diagnostic capability continues to proceed. The sensors and actuators used in the future will probably become "intelligent", containing their own self-diagnostic capabilities. This will offer greater self-diagnostic sophistication and also free up the central processing unit to perform its main control tasks more efficiently. The article following this one will explain the proper procedures for using the scan tool to interface with the GM data stream. As you will see, the data stream offers a wealth of information and options far beyond the fault code information.

Using The Scan Tool

GM EFI engine controllers have one very convenient feature in common with each other. All of the GM ECMs contain a serial transmitter within the ECM which can transmit to another computer various computer "words" which represent input data being received by the ECM and output commands being issued by the ECM. The scan tool is a mini-computer designed to interface with the GM data stream. The original scanners were usually only interfaceable with one specific ECM type. The next generation of scan tools were called multi-scanners because they were re-programmable to interface with more than one type of ECM. They also featured dual display read-outs and two different data stream words could be displayed at one time. Today's scan tool features almost complete compatibility with any GM ECM which has a data stream. The displays have been expanded to multi-line displays which will display many more data stream words simultaneously. Some tools also have LED panel indicators which add additional simultaneous monitoring capability.

The skilled GM EFI technician today has learned what benefits the scan tool offers and how to use them. The skilled technician has also learned what the scan tool won't do. When the display says "NO DATA", the technician must know what to do and how to read the circuit information without the use of the scan tool.

The scanner is definitely a great tool and the GM technician of today wouldn't be without one. It offers very fast access to a large amount of information which can in many cases determine the cause and extent of the driveability problem. One of the problems with the use of the scanner is the fact that the information is not fresh enough due to data stream communication rate or even data stream error. Also, in some cases, the scanner will modify the ECM behavior and actually cover up or disguise the problem or problems.

To access the data stream, the tool user needs only to properly set up the scanner's parameters, plug the test lead into the ALDL, and turn the ignition key on. The tool can also be used during a road test because the tool is designed to be vehicle battery powered through the cigarette lighter socket. As the data stream communicates information to the scanner, the technician can select different scan tool test modes which will determine which computer words are being displayed. The readings displayed will represent fault codes, sensor input values or ECM output commands. These readings are in many cases converted values. A good example would be the engine coolant temperature reading. The scanner will display this information as temperature, either Celsius or Fahrenheit, when the ECM is actually receiving the information from the circuit as an analog voltage. Understanding what the circuit values are, what they should be, or what variations from the norm can mean is directly related to the technician's ability to understand the circuits and the system they comprise.

Using The Scan Tool

Cont.

A properly trained technician can spot circuit or subsystem abnormalities by assessing the data stream values and comparing them to the known good, expected parameters. The properly trained technician can also perform the same task without the aid of the scan tool or data stream access. System and circuit knowledge is absolutely essential to the success of the driveability technician.

The only variation to the scan tool operation concerning GM EFI vehicles which might affect the tool's operation is the option of the diagnostic state in which the scanner is operated. Most scanners are programmed to automatically set up in the diagnostic state known as ROAD TEST, OPEN or NORMAL mode. This state is the diagnostic state which should not cause the ECM to modify any of its commands when the scanner is hooked up. Some of the early GM EFI engine controllers did not transmit the full data stream when in this mode. They were designed to transmit the full word stream only when the ECM diagnostic state called the ALCL STATE was used. This state requires the scanner to lower the diagnostic request line voltage to approximately 2.5 volts and the ECM is designed to modify its output command behavior to some extent.

The other diagnostic states for GM EFI engine controllers are FIELD SERVICE STATE and BACK-UP FUEL STATE. Each of these states have specific diagnostic purposes and the technician should also be very familiar with these special modes. The scanner does not pick up any data in these modes. It is a convenient way to instruct the ECM to enter these diagnostic states.

In summary, the scan tool is a very dynamic tester which will allow the technician a lot of flexibility in testing GM EFI systems. Learn to utilize the tool to its fullest capability. Also, learn the system and its circuits so the scanner becomes an outstanding verification tool to speed up information access.

ALDL or ALCL Connector

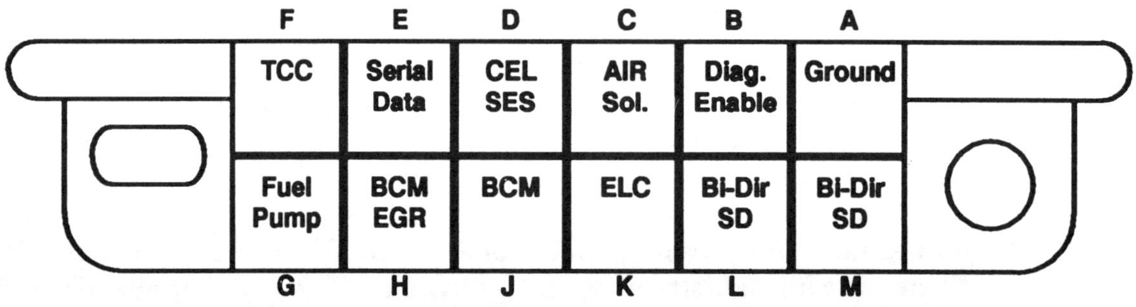

ALDL—Abbreviation for Assembly Line Diagnostic Link

ALCL—Abbreviation for Assembly Line Connector Link or Assembly Line Communication Link

Different publications will use either term. Both are correct.

Every GM fuel injection vehicle will have one of these connectors. Cadillac DFI has an ALDL, also, but the serial data is not directed to the ALDL. Hence, your scan tool does not read the inputs and outputs as it does on all other GM vehicles.

With nothing connected to the ALDL, the ECM "sees" 5 volts on the Diagnostic Enable line terminating at "B". Jumpering "B" to "A" will drop the "B" line voltage to 0 volts and this signals the computer to initiate a series of routines which are programmed into the self-diagnostic capacity section. This is the procedure called "Diagnostic Circuit Check." This state is applied with the "Key On, Engine Off."

1. The ECM will flash the CEL signaling any fault codes which are present or in long term memory.

2. The ECM also actuates most of its 12 volt solenoid actuators allowing the technician to test each circuit.

ALDL or ALCL Connector

Cont.

3. The ECM also issues extend-retract counts to the IAC, again allowing the technician to make circuit and actuator checks.

So, jumpering terminal "B" to "A", or dropping terminal "B" voltage to 0 volts is called the "Zero Ohms Mode" or the "Diagnostic Mode" or the "Diagnostic Circuit Check" if the line is jumpered with the "Key On, Engine Off."

If the terminals "A" and "B" are jumpered together with the "Key On, Engine Running," then the ECM acknowledges this as a request for another Diagnostic State. This state is called "Field Service Mode." Since the ECM is seeing a distributor reference signal, yet the diagnostic enable line is pulled to 0 volts, the ECM does not flash the Check Engine light in sequence to issue codes. Instead, if the ECM is in open loop, it will flash two times per second. If the engine is being controlled in closed loop, the light will flash one time per second. The amount of "Light On" time will inform the technician how rich or lean the engine is. If the light is "On" 50% and "Off" 50%, the ECM is sending neither a "Rich" or "Lean" command. If the light is on more than off, then the ECM is detecting enleanment and conversely, it is sending enrichment commands. Inversely, if the light is off more than on during each one second period, then the ECM is detecting enrichment and it is sending enleanment commands.

If a scanner tool is connected to the ALDL, in its normal operating state, the scanner places a resistor across the "B" and "A" terminal of about 10k ohms. This then drops the voltage that the ECM "sees" to about 2.5V and this initiates a different response from the ECM. Now the ECM begins issuing it's serial data at the designated terminal. If it is an early system, it will issue that data to terminal "E". If it is the later P-4 system, it could possible issue that data to terminal "L" or "M" or both, depending on the protocol for that system. That series of "words" can only be received and interpreted by another device such as a scan tool. Those electronic signals can be recorded and played back to an interpreting device. This mode of diagnostics is called the "special" or "ALDL" mode. It is important for the technician to understand that in this mode, the ECM will issue some very useful information if the technician understands it. The car cannot be properly tested in this mode for all driving conditions because the scanner causes the ECM to command an idle speed of 1000 RPM. Vehicles with ESC may even severely knock in this mode due to ESC disablement. With the scanner plugged in, the ECM bypasses timer sequences and accurate evaluation on many subsystems is now modified. This state is called the ALDL or ALCL state.

If the scanner is connected to the ALDL and the Road Test state is selected, the voltage on the Diagnostic Enable terminal is not dropped to 2.5 volts. The resistance applied between ALDL terminals "A" and "B" is very high and therefore the voltage

ALDL or ALCL Connector

Cont.

drop is typically less than 1 volt. In some cases, the voltage will remain at or near 5.0 volts on the diagnostic terminal. This informs the ECM that the serial data line is being monitored for ECM inputs and outputs while test driving or road testing the vehicle. This mode should prevent the scanner from causing ECM modifications, such as idle, timing or complete time out changes which would affect the diagnosis. The condition to watch for is to note whether or not the full data stream is available. Some vehicles will not issue the data stream or will blank out certain "words" in the data stream in this mode. If the data stream is not available, or is only partially available, the technician may have to change to the 10k state in order to gain access to the information. Most early scanner software automatically operated in the 10k state and had to be manually instructed to go to the Road Test mode. Most late model scanner software automatically defaults to the Road Test mode when used.

If the ECM detects certain internal malfunctions which might prevent the vehicle from being driven to the repair shop, the ECM activates an internal "CalPac" which substitutes special "Limp Home" fuel commands to the injectors. The scan tool can be used to initiate the state. By applying 3.9k ohms of resistance across the A-B terminals, the diagnostic enable line voltage will drop to about 1.0V. This will cause the computer to default to the Fuel Back-Up State. This state is sometimes called the 3.9k mode.

In summary, reading data on a GM EFI system with a scan tool entails merely identifying the system, receiving and reading the emitted data in a simple function called "Data Stream Monitoring". The only variation to this is to alter the amount of resistance placed across the diagnostic terminals by the scanner which will cause the ECM to enter any one of these four diagnostic states:

1. State 6 – Field Service
2. State 7 – Back-Up Fuel
3. State 8 – ALCL, 10k or Special
4. State 9 – Open or Road Test

Learn to utilize the scanner in these four states to gain access to the wealth of diagnostic information divulged by the ECM.

ALDL or ALCL Connector

Cont.

ALDL Circuits "A-B" and "E"

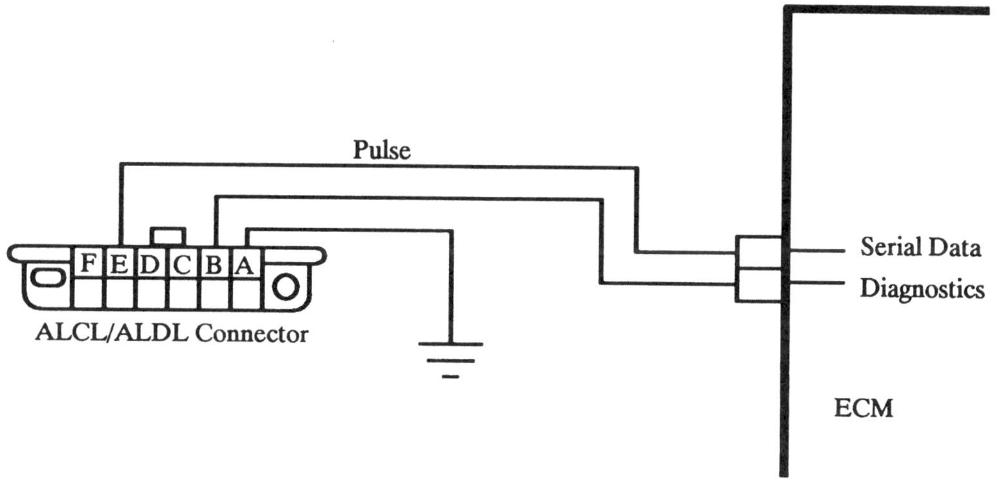

Approximate "B" Line Voltage	Diagnostic State
5.0V	Open/Normal – 0 Ohms (Road Test)
2.5V	ALCL – 10k Ohms
1.0V	Backup – 3.9k Ohms
0.0V	Field Service Mode (Engine Running) or Diagnostic Circuit Check – 0 Ohms (Key On/Engine Off)

Approximate "E" Line Voltage	
2.0 – 5.0V	Digital Serial Data Transmission

ALDL or ALCL Connector

Cont.

ALDL Circuit "C"

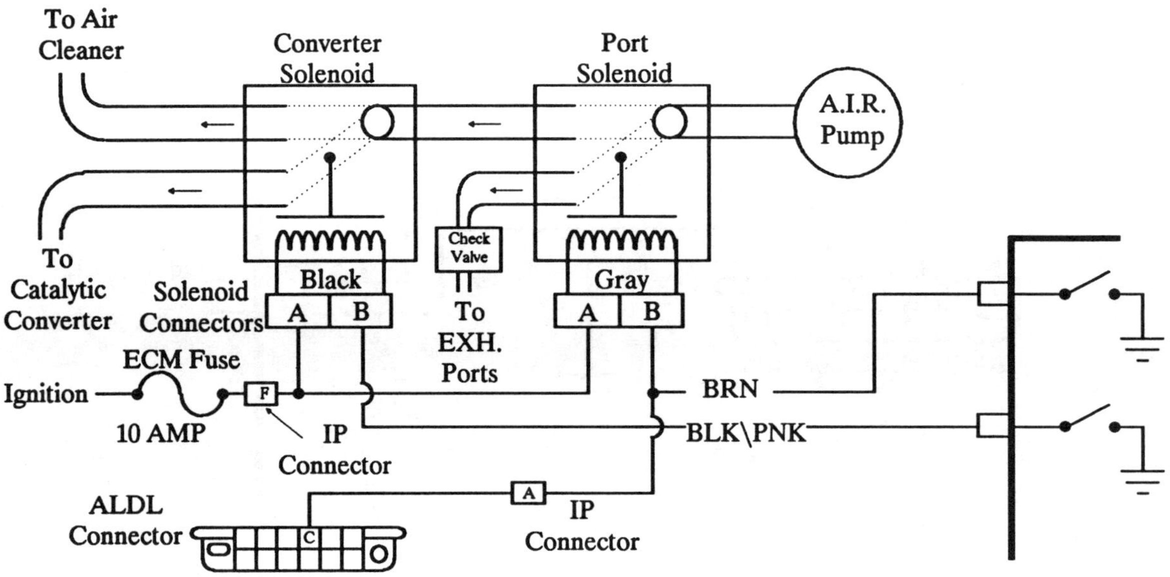

When the ECM has the driver transistor "ON", the solenoid circuit is completed and the voltage on the solenoid's ground circuit is 0.0V. When the driver transistor is "OFF" the voltage is battery voltage. Pressing the zero key on most scanners overrides the ECM and turns the driver transistor "ON" for about 10 seconds, allowing the technician to verify operation of the circuit, solenoid and air valve. Observing the oxygen sensor voltage will indicate whether the air was directed upstream during that test. The oxygen sensor's ability to detect abundant oxygen and its reaction time can be verified in this manner.

Approximate "C" Line Voltage	Condition
Battery V	Key On/Engine Off
0.0V	Key On/Engine Running, Open Loop, Medium Throttle
Battery V	Key On/Engine Running, WOT
Battery V	Key On/Engine Running, Closed Loop

ALDL or ALCL Connector

Cont.

ALDL Circuit "D"

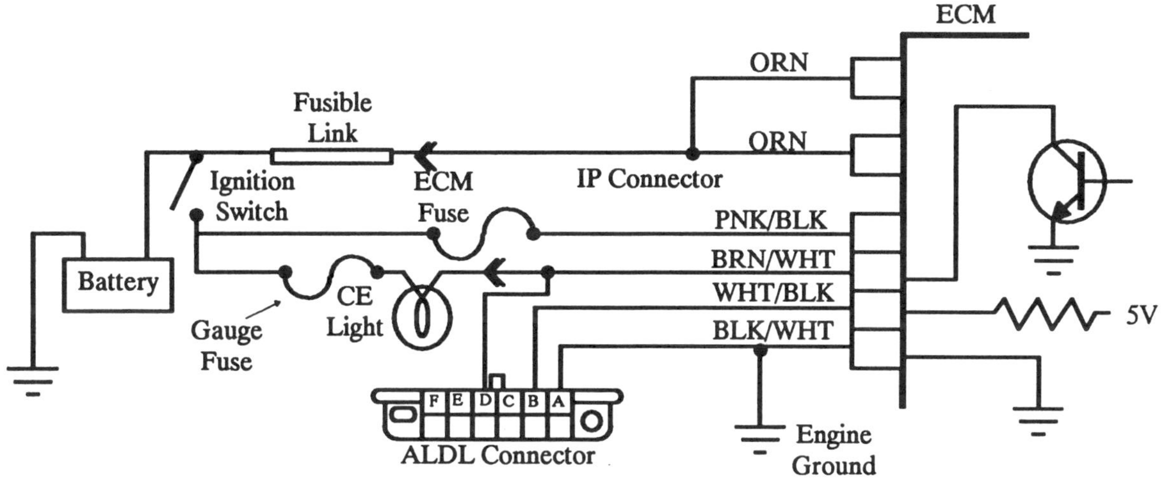

Power is supplied to the Check Engine Lamp through the ignition circuit. The ground side is controlled by the ECM. When the CEL driver transistor is "ON", the circuit is completed and the lamp should illuminate. Voltage on the ground side of the lamp should be 0.0V. When the driver transistor is "OFF", the lamp should be out and the ground circuit voltage will be battery voltage.

Approximate "D" Line Voltage	Condition
0.0V	Key On, Lamp On
Battery V	Key On/Engine Running, Lamp Off (No Hard Fault Present)

ALDL or ALCL Connector

Cont.

ALDL Circuit "G"

Chevrolet Engines

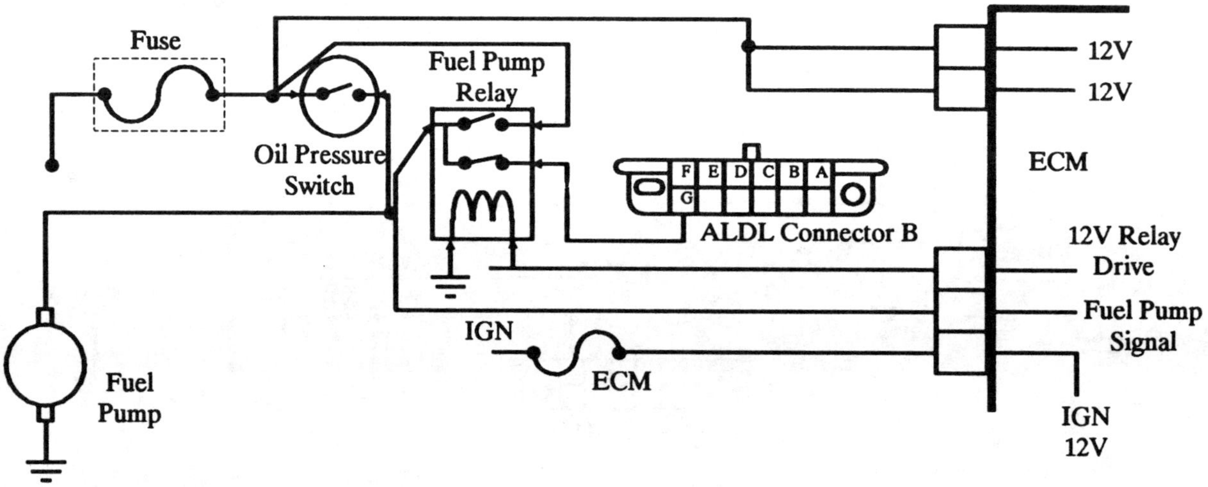

On some GM vehicles the fuel pump relay circuit has an alternate fuel pump power route which terminates at the ALDL terminal "G". This will allow the technician to apply battery voltage to ALDL terminal "G" and activate the fuel pump continuously. This procedure can be used to test the fuel pump and fuel pump flow. Inversely, the voltage present at terminal "G" can be used to indicate whether the fuel pump is "ON" or "OFF".

Non-Chevrolet Engines

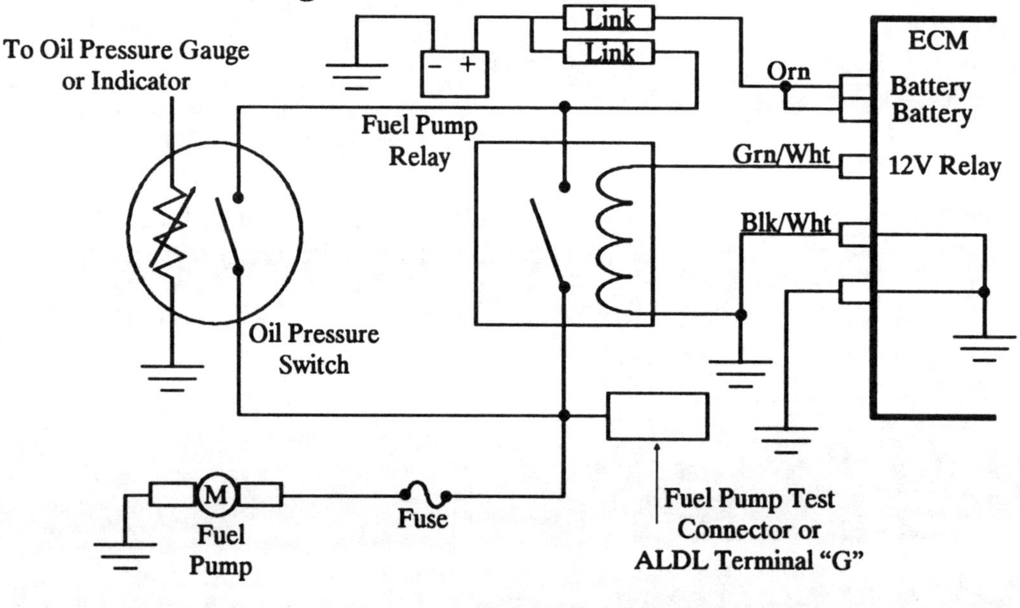

ALDL or ALCL Connector

Cont.

ALDL Circuit "M" or "L"

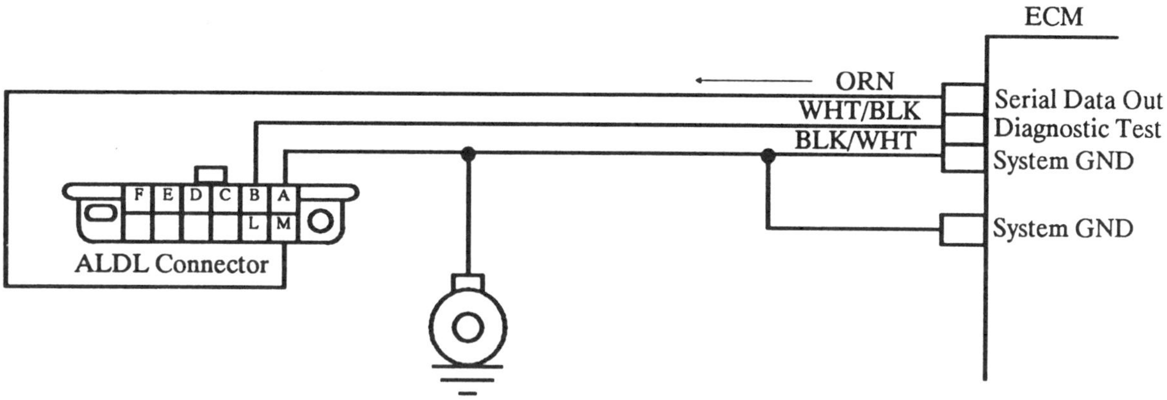

On many later model GM vehicles, the serial data issued by the ECM is directed to ALDL terminal "M" instead of "E". Some publications will call this data line the "twisted pair" data circuit referring to the "L-M" circuit. This data line operates differently than the old line, plus the data issue rate has been significantly increased. These terminals are called "bi-directional" in that computer generated information can be sent into as well as out of the ECM. This will allow GM to make internal software changes within the ECM instead of having to make field hardware changes. Also, diagnostic testing will utilize this feature to activate internal ECM subsystems to make temporary calibration changes to verify problems or solve engine performance problems. If the ECM has a connection to this circuit, then the data stream baud rate is the newer high speed data rate. Only a limited number of GM vehicles have this feature.

General Fault Code Definitions

Code	Definition
12 -	No Reference Pulses to ECM
13 -	O^2 Sensor Circuit – Will Set If: At least two minutes after engine start. O^2 signal voltage steady between .35 and .55 volts for more than one minute. Throttle position sensor signal above 6% (about 1200 RPM).
14 -	Coolant Sensor Signal Low – Will Set If: Signal voltage indicates the coolant temperature is above 135° C (275° F) for more than two seconds on some engines, ten seconds on other engines.
15 -	Coolant Sensor Signal High – Will Set If: Signal voltage indicates a coolant temperature less than –35° C (–24° F) for four seconds, and time since engine start is more than 30-60 seconds.
16 -	High battery voltage
	Or
16 -	DIS fault line open/shorted
17 -	RPM signal problem
21 -	TPS Signal Voltage High – Will Set If: TPS signal voltage is greater than 2.5 volts for two seconds on some engines, 3.5 volts for five seconds on other engines. Engine speed is less than 1600 RPM. MAP is equal to no load which is less than 60 KPA (9 PSI) on some engines, 30 KPA (4.5 PSI) on other engines.
22 -	TPS Signal Voltage Low – Will Set If: the TPS signal voltage is less than 2 volts for two to five seconds. engine speed is less than 1200-1600 RPM.
	Or
22 -	Fuel cutoff relay circuit open/grounded

General Fault Code Definitions (Cont.)

Code	Definition
23 -	Manifold Air Temperature Low
	Or
23 -	TPS Error
24 -	Vehicle Speed Sensor – Will Set If : Circuit 437 voltage is constant, no pulsing, for a period of four to ten seconds. Engine speed is between 1500 and 440 RPM. Park/neutral switch indicates the transmission is in the drive range. Low MAP (high vacuum) indicating engine deceleration.
25 -	Manifold Air Temperature High
	Or
25 -	Vacuum Switching Valve Circuit Open/Shorted
	Or
25 -	ATS sensor high
26 -	Quad Driver Module (QDM)
27 -	2nd Gear Switch
28 -	3rd Gear Switch
29 -	4th Gear Switch
31 -	MAP Low
	Or
31 -	Fuel Injector (1989)
	Or

General Fault Code Definitions (Cont.)

Code	Definition
31 -	Park/Neutral Switch
	Or
31 -	Cam diagnostic
	Or
31 -	Governor malfunction
	Or
31 -	Wastegate overboost
	Or
31 -	Wastegate Electrical Signal Open/Grounded
32 -	BARO Circuit Failure
	Or
32 -	EGR Failure – Will Set If: The EGR diagnostic switch is closed during start-up. "Check Engine" light will come on within 40 seconds and will stay on until the switch is opened or the engine is turned off. The EGR diagnostic switch was open under the following conditions: Coolant temperature greater than 90° C (194° F). EGR duty cycle commanded by the ECM is greater than 50% Above conditions met for approximately four minutes. "Check Engine" light will remain ON only if all these conditions are met.
	Or
32 -	EGR/EVRV

General Fault Code Definitions (Cont.)

Code	Definition
33 -	MAF Frequency or Voltage High – Will Set If: Low engine RPM. TPS at idle. Airflow or frequency is greater than 20 grams per second. Time lapse of 3 to 5 seconds.
	Or
33 -	MAP Sensor Signal Voltage High – Will Set If: Voltage signal is too high for more than five to eight seconds. TPS voltage indicates the throttle is closed. Engine misfire or unstable idle may also set code 33. Vehicle is stopped.
34 -	MAF Frequency or Voltage High – Will Set If: Engine is running over 1400 RPM. TPS over 2.5 volts. MAF sensor is disconnected. Airflow is less than 2 grams per second (2.8L). Airflow is less than 10 grams per second (3.0l & 3.8L).
	Or
34 -	MAP Sensor Signal Voltage Low – Will Set If: The ignition is on. The signal voltage is too low.
35 -	IAC Problem 100 RPM above or below desired RPM for 45 seconds.
36 -	Mass Air Flow/Burn-off circuit problem.
	Or
36 -	Transmission shift problem
	Or
36 -	DIS fault indicating extra/miss EST
38 -	Brake Switch Circuit

General Fault Code Definitions (Cont.)

Code	Definition
39 -	TCC Circuit
41 -	Cam Sensor Failure
	Or
41 -	Cylinder Select Error or no reference pulses with engine running.
	Or
41 -	Tach input error
42 -	Electronic Spark Timing – Will Set If: The engine is running for approximately one minute. Open or short-to-ground on the EST or bypass circuit.
	Or
42 -	Fuel cutoff relay/circuit shorted
43 -	Electronic Spark Timing – Will Set If: Low voltage at ECM's ESC terminal for more than four seconds.
44 -	Lean Exhaust Indication – Will Set If: The oxygen sensor voltage at the ECM on circuit 412 remains below .2 volts for eight seconds or longer. Time since engine start is one to two minutes or longer.
45 -	Rich Exhaust Indication – Will Set If: The oxygen sensor voltage at the ECM on circuit 412 remains above .7 volt for 30 to 60 seconds or longer. Time since engine start is one minute or longer.
46 -	VATS failure
	Or
46 -	Power steering switch

General Fault Code Definitions (Cont.)

Code	Definition
47 -	UART link
48 -	Misfire
49 -	Vacuum leak
51 -	PROM or ECM
	Or
51 -	Check Sum error
52 -	Fuel Calpac missing
	Or
52 -	QDM fault
	Or
52 -	A/D error
	Or
52 -	Oil temperature sensor low
53 -	Over voltage condition
	Or
53 -	EGR
	Or
53 -	V reference error
	Or
53 -	Vehicle Anti-Theft System (VATS) diagnostics

General Fault Code Definitions (Cont.)

Code	Definition
54 -	Low Fuel Pump Voltage or Fuel Pump Relay
	Or
54 -	QDM component failure
55 -	ECM failure
	Or
55 -	Serial bus error
	Or
55 -	SAD error
	Or
55 -	Fuel lean malfunction
56 -	Corrosivity/Add coolant
	Or
56 -	Port throttle system vacuum sensor
58 -	Vehicle Anti-Theft System (VATS) problem
61 -	Oxygen Sensor Degraded
	Or
61 -	Port throttle system error
	Or
61 -	Cruise vent solenoid circuit

General Fault Code Definitions (Cont.)

Code	Definition
62 -	Gear switch inputs diagnostics
	Or
62 -	Oil temperature sensor high
	Or
62 -	Cruise vacuum solenoid circuit
63 -	MAP Voltage High - 2.8L with MAP/MAF
	Or
63 -	Small EGR Failure
	Or
63 -	Right oxygen sensor failed
64 -	MAP Voltage Low - 2.8L with MAP/MAF
	Or
64 -	Medium EGR Failure
	Or
64 -	Right oxygen sensor lean
65 -	Large EGR Failure
	Or
65 -	Injector peak/hold diagnostic
	Or
65 -	Right oxygen sensor rich
	Or

General Fault Code Definitions (Cont.)

Code	Definition
65 -	Cruise servo position sensor
66 -	A/C pressure transducer fault
	Or
66 -	ECM reset
67 -	Cruise switches circuit
68 -	Cruise system circuit
69 -	A/C pressure switch circuit
88 -	ECM reset

Pin Point Circuit Testing

Now let's discuss the steps involved in pin pointing the cause or causes of a system problem. After the technician has entered the on-board diagnostic system and tested the system for the presence of any coded faults, the next steps will involve testing the individual circuit or circuits which could possible cause the fault code or codes. The GM engine performance service diagnostic literature contains the appropriate procedures for locating these circuit faults. The information is almost always displayed as a diagnostic "tree". Many of the steps utilized in following the diagnostic "tree" are designed to "prove" the circuit wiring. The "tree" almost always directs the technician to "backprobe" the ECM harness connector to verify the presence or absence of the expected circuit voltage. These "backprobing" steps can be very confusing and there is always the possibility that the pin or terminal can be missed and a wrong diagnostic conclusion can be made. The professional electronic technician, when confronted with a large number of circuits to test, will usually use a "breakout" style interface when one is available. Ford technicians utilize a Ford Breakout Box for testing since Ford controllers do not have a serial data stream like GM.

The tester illustrated below is manufactured as a professional breakout style interface for GM ECM connectors. It has been field proven to be a very successful tool for locating circuit faults on GM computer controlled vehicles. The interface allows the technician to "tap" into all of the ECM circuits simultaneously, allowing the technician to access the incoming and outgoing signals with the testers of his choice. The real key advantage of this style of testing is the fact that the readings taken will be "live" circuit readings, not delayed or averaged serial data stream readings.

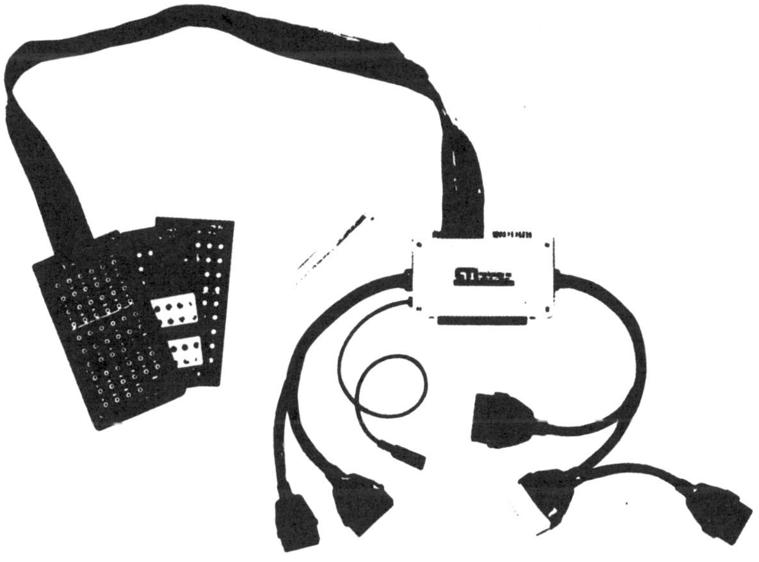

Model 333 Test Probe Interface – Made by Concept Technologies

Pin Point Circuit Testing

Cont.

By connecting the test probe interface between the ECM and the vehicle harness, safe and sure test access to each ECM input and output circuit can be gained. Since GM's ECMs are usually located inside the vehicle passenger compartment, the interface can be placed on the front seat of the vehicle and measurements can be made while the vehicle is being driven or operated under actual road-load conditions. This has proven to be a superior method of locating circuit faults or ECM behavior problems which are intermittent or which occur only under road-load conditions. Accessing the circuits with a test probe interface offers the technician the ability to stall test or road test the circuits with the test instruments of his choice. Remember, all of the incoming input signals and all of the outgoing commands terminate at the ECM harness connectors. "Real time" circuit testing is by far the best test method of proving the circuits and the devices contained in each circuit.

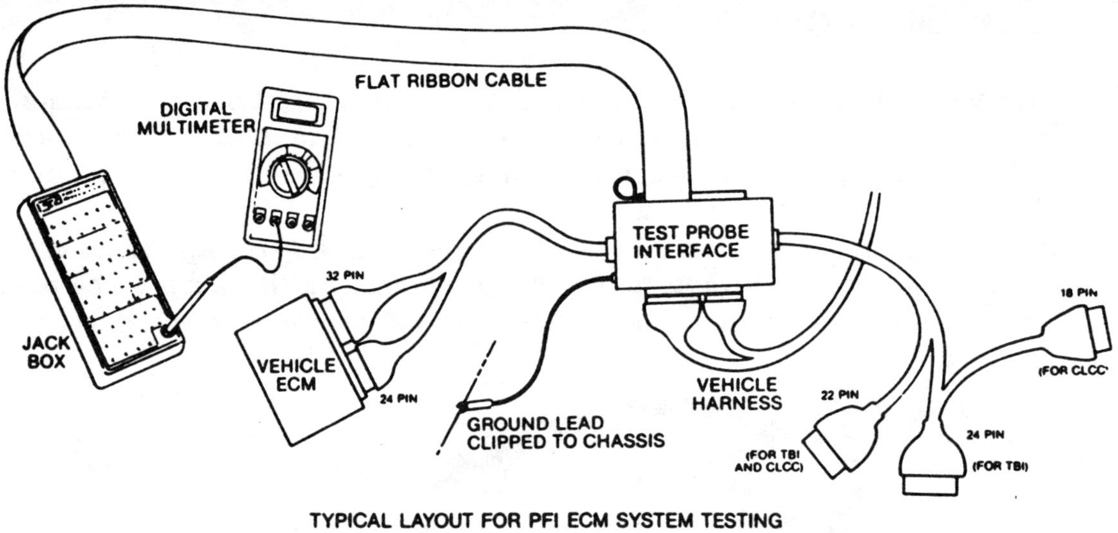

TYPICAL LAYOUT FOR PFI ECM SYSTEM TESTING

ECM Parameters – TBI

INPUT

- Coolant Temperature Sensor
- MAP Sensor
- Engine Crank
- Detonation Sensor
- Exhaust Oxygen - O^2
- Crankshaft Position & RPM
- Throttle Position Sensor
- Air Temperature Sensor
- EGR Vacuum Switch 3rd or 4th Gear Switch
- Park/Neutral Switch
- System Voltage
- Power Steering Load
- Cruise Control Status
- Vehicle Speed Sensor
- A/C Request

→ **Electronic Control Module** →

OUTPUT

- Transmission Converter Clutch (Shift Light)
- Fuel Pump Relay
- Air Conditioning Relay
- Engine Cooling Fan Relay
- Canister Purge Solenoid
- Exhaust Gas Recirculation Solenoid
- Injector(s)
- Electronic Spark Timing
- Idle Air Control
- "Service Engine Soon" Light
- Serial Data Line
- Diagnostics
- Air System Valve or Valves

Note: *Not all systems used on all vehicles.*

Many of these sensors were used on CCC Carbureted vehicles. They still function in a similar manner, yet in many cases they have been re-assigned new priorities in the protocol. Because they have new roles to play, the conditions needed to set fault codes have been changed dramatically. It is important that the technician refamiliarize himself with those new conditions.

But remember, the essential task of allowing the computer to intelligently control fuel management, spark management, air management, transmission control and some of the emission control devices is still the ultimate goal of these systems.

Also, please be constantly aware that there are still numerous engine system malfunctions that will cause serious driveability problems without setting any trouble codes. The technician must always check for the presence of those types of failures before he condemns the computer or its related input or output devices.

ECM Parameters - PFI

INPUT

- Coolant Temperature Sensor
- Detonation Sensor
- Exhaust Oxygen - O^2
- Camshaft Position Sensor
- Crankshaft Position & RPM
- Throttle Position Sensor
- MAF Sensor
- Air Temperature Sensor
- MAP Sensor
- BARO Sensor
- EGR Vacuum Sensor
- 2nd, 3rd or 4th Gear Switch – Park/Neutral Switch
- System Voltage
- Power Steering Load
- BCM 800 Circuit
- Cruise Control Status
- Vehicle Speed Sensor
- A/C Request

→ **Electronic Control Module** →

OUTPUT

- Transmission Converter Clutch (Shift Light)
- Fuel Pump Relay
- Air Conditioning Relay
- Engine Cooling Fan Relay
- Canister Purge Solenoid
- Exhaust Gas Recirculation Solenoid(s)
- Injectors
- Idle Air Control
- Electronic Spark Timing
- Electronic Spark Control
- "Service Engine Soon" Light
- Serial Data Line
- Diagnostics
- Air System Valve or Valves
- Early Fuel Evaporation
- Turbo Wastegate

Note: *Not all systems used on all vehicles.*

The only significant changes in sensor inputs for PFI compared to TBI was the conversion from using a MAP (Manifold Pressure control) to the utilization of a MAF (Air Flow control). GM's mass airflow sensors are quite sophisticated in design and operation. You can study them in greater detail in the section on Mass Airflow Sensors. On the early 2.8L V6 Pontiac Fiero, the main ECM input was a MAP sensor. That system was computing fuel demand based on engine load as seen by ECM through manifold pressure. The new Oldsmobile Quad 4 engine also uses a MAP Sensor as its main sensor input for fuel control. Most other GM engines use the MAF as the main input for the ECM to calculate fuel requirements. On the very latest models, certain engines will be equipped with both a MAF and a MAP. On these units, the MAF is the engine's main fuel requirement sensor. But in the event of a MAF failure, the ECM is programmed to continue to deliver fuel based on the backup input of MAP.

ECM Parameters - PFI

Cont.

The only output device change from TBI to PFI involves the ECM controlling multiple injectors instead of just one or two. On most engines, the injectors are group or "gang" fired and so the ECM has no special task to perform. However, on the engines with sequential firing fuel injection, the technician must take notice that there will be an individual circuit for each injector.

On either TBI or PFI fuel injected engines, the ECM has total responsibility to administer the fuel into the engine. There is no mechanical device such as an accelerator pump to add or deliver needed fuel. Therefore, it is important that the ECM be effective in delivering accurate amounts of fuel needed for all engine modes. The calculations necessary to accomplish this are much more complex and all of the inputs needed by the computer are each very important in their own right. Thorough examination of all inputs is always an absolute necessity.

The most significant change in sensor inputs for PFI as compared to TBI was the use of the Mass Airflow Sensor or MAF. Most of the early GM PFI engines used a MAF sensor to compute engine fuel requirements. Most of those engines did not have a MAP sensor, a few of them did. In 1989, the Chevrolet-Pontiac-Canada Group engines discontinued using MAF sensors and went to the speed density type of fuel control system which required MAP sensors instead of MAF sensors. Buick still uses the MAF sensors on the V6 engines. Always identify the system and the sensors utilized to prevent a misunderstood code or a possible diagnostic error.

The most significant output device change on PFI as compared to TBI is the use of more output drivers for injector control. The TBI engines required only one or two output injector drives. The group or gang fired PFI engines will also only require two injector drivers. The sequential PFI engines will require up to eight output drivers.

Remember, on EFI engines, either TBI or PFI, the ECM has total control of the admission of fuel into the engine. There is no mechanical device such as an accelerator pump or choke to add needed fuel. A few GM engines do use a cold start injector, but that injector is operated by a thermal time switch, not the ECM. Therefore, it is necessary for the ECM to have all the needed input information to properly calculate the various control signals to deliver the right amount of fuel for all of the engine operating modes. The calculations required to accomplish this task are more complex for PFI than for TBI. Each input signal is important and thorough testing of all of the necessary signals is very important.

ECM Circuits

Inputs

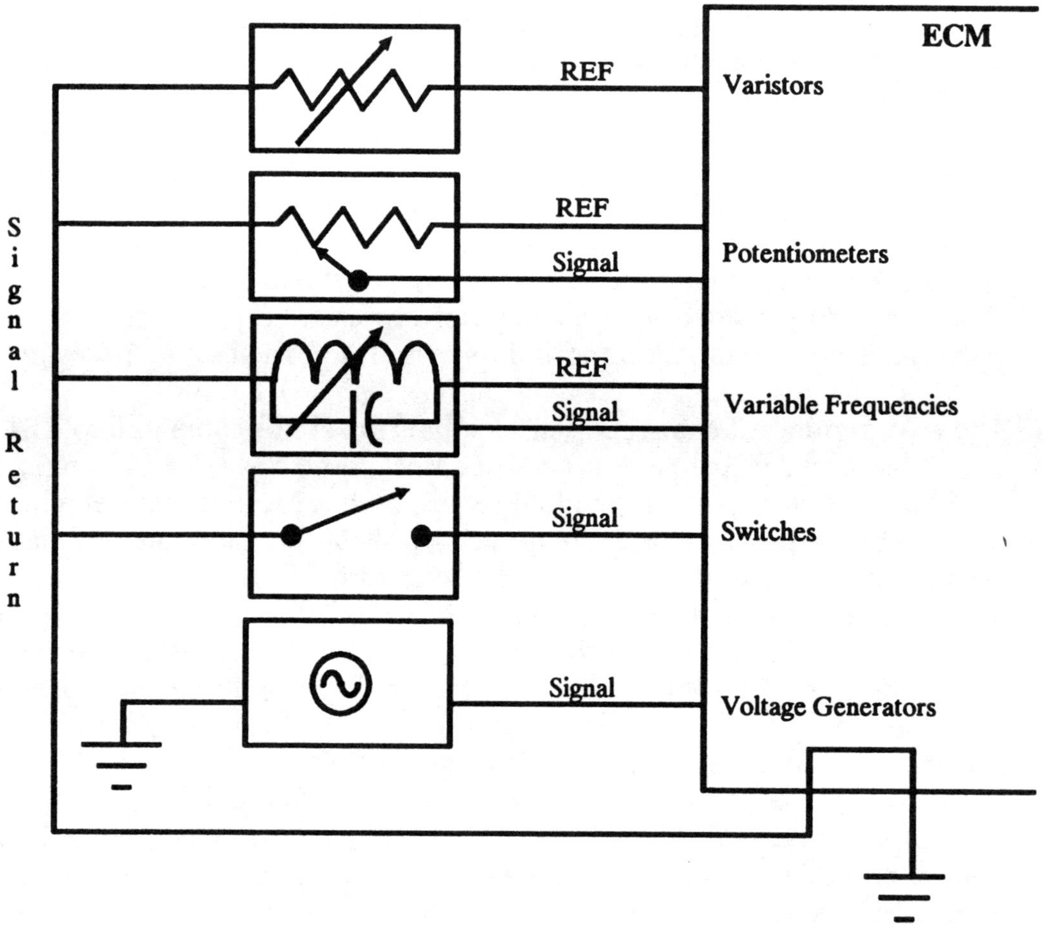

All of the input signals to the ECM are simple device generated or modified signals. These circuits are very high resolution, low current circuits and accurate measurements should always be taken to safeguard the device and the ECM. The input signals you detect through the data stream will in many cases be "converted" readings. For instance, the ECM receives the MAF signal as a variable voltage or frequency. It displays that signal as "grams per second" air flow. This signal conversion is easy to understand if you understand the circuit and input signal device.

ECM Circuits (Cont.)

Outputs

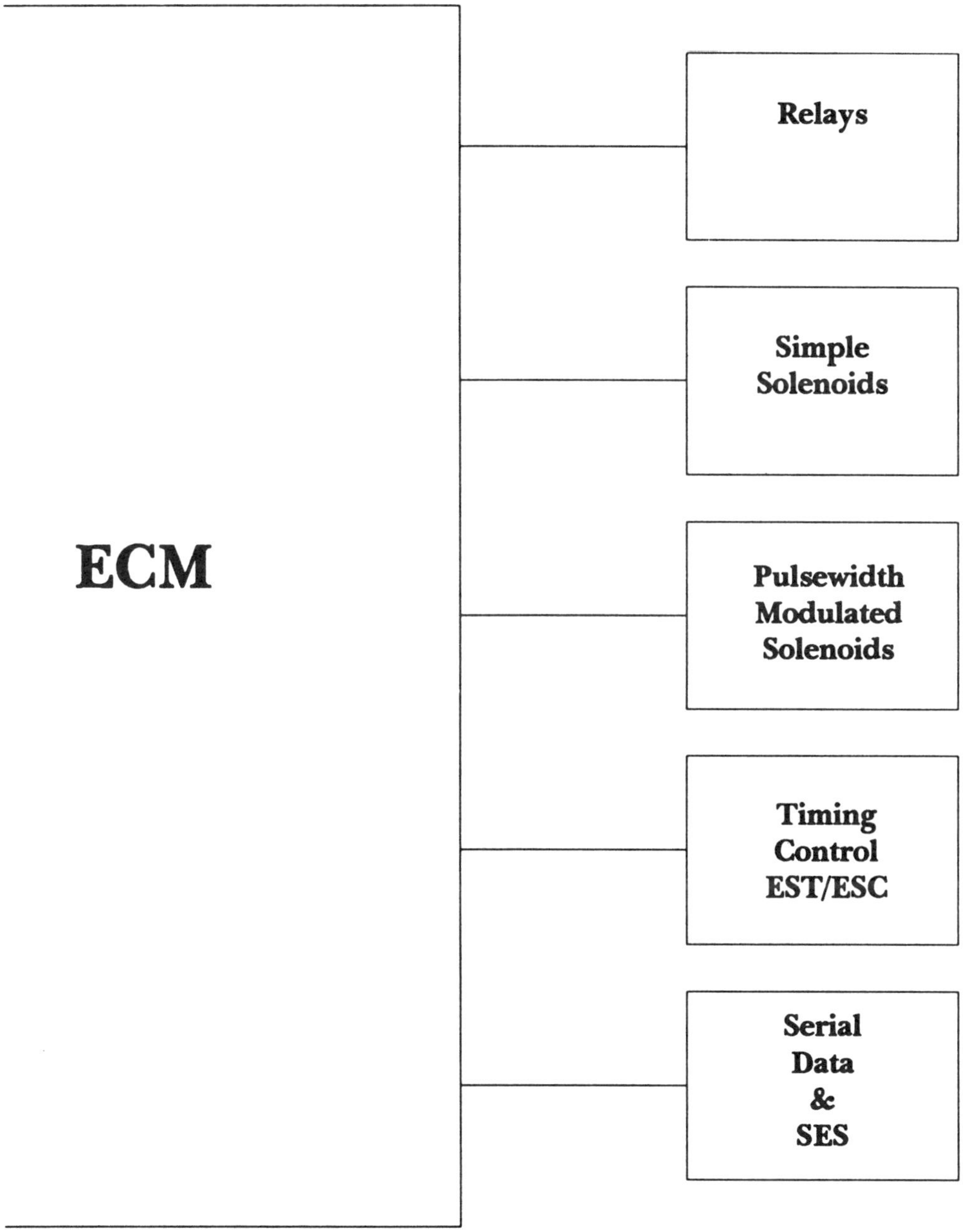

ECM

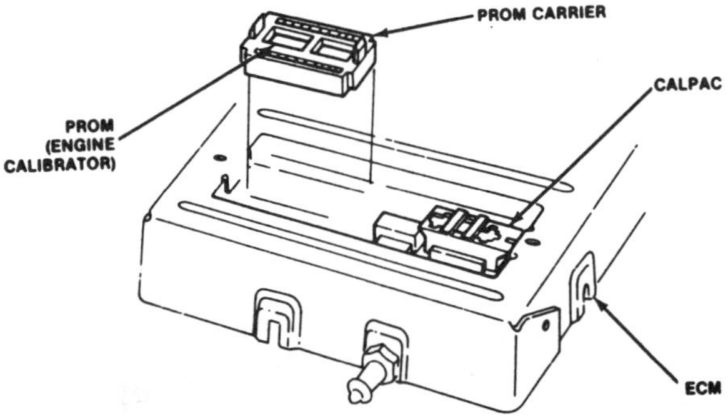

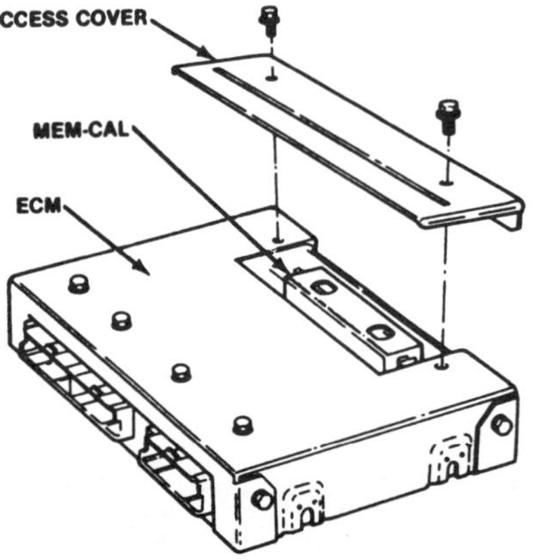

Courtesy of AC-Delco

What Is It?

The Electronic Control Module, or ECM, is a high speed microprocessor which is capable of receiving inputs from various vehicle sensors, processing those inputs and issuing various commands to the controlled outputs. The ECM on most GM EFI vehicles is located inside the passenger compartment. The ECM consists of essentially two and possibly three replaceable parts. The controller, the PROM and in some cases a CALPAC. Many late model GM ECMs now contain a MEM-CAL which contains the PROM and the CALPAC. The MEM-CAL assembly can also contain the ESC module. The microprocessor used in the ECM has been continually upgraded and today's ECMs usually contain a Motorola 68HC11. The wiring harness which connects the ECM to the system inputs and outputs will utilize either edgeboard style connectors or header style pin connectors. New replacement ECMs are quite expensive. For most applications, a remanufactured ECM is available at a greatly reduced cost. The PROMs, CALPACs and MEM-CALs are only available as new replacement units.

What's Its Job?

The ECM is designed to control the injectors, the idle control device, the timing management system, and various other subsystems such as emissions, cruise control or transmission. The role the ECM plays on different vehicles will vary depending on the application. All GM ECMs have a self-diagnostic subsystem with memory. The self-diagnostic subsystem is designed to continuously monitor the system and to operate the CHECK ENGINE or SERVICE ENGINE SOON lamp to inform the vehicle operator if the system has detected any system faults.

ECM

Cont.

Circuit Fault Codes:

The ECM can detect certain problems within itself and set fault codes such as: Code 51, Code 52, Code 53, Code 54, Code 55 or Code 88. In actuality, any fault code can signal an ECM failure since it is possible for an input or output circuit to be healthy, yet set a fault code due to ECM error.

What To Watch For?

The ECMs used on GM EFI engines have proven to be highly reliable. The largest percentage of replaced ECMs are found, when tested, to be in good operating order. The largest percentage of failed ECMs reveal that the unit was damaged by the handler. Electro-static discharge from hair or clothing seems to be a common culprit. GM ECMs are not hermetically sealed. If the ECM is exposed to the elements, the corrosion can damage the ECM or its connections. Welding or boost starting an ECM equipped vehicle can also cause damage. Computers are sensitive devices so use caution when working around them or on them.

ECM Circuit Troubleshooting

To test all of the circuits with a GM ECM would require a very sophisticated array of system emulators and monitors. There is a large number of circuits and devices contained in the ECM and these vary greatly from one ECM to another.

When following the circuit tests for the various sensors and actuators, the ECM's role or job in each case should be considered. Many of the tests you will perform are actually signal change emulations to test the ECM's ability to monitor the change and issue a command.

The easiest way to examine an ECM and determine whether it is healthy or not is to define its role as basically as possible and perform a few basic tests to verify its ability to fill that role.

TEST	DETERMINATION
Power Supplied	The ECM recieves B+ at several terminals. Some terminals should be HOT at all times. Some terminals are HOT only when the key is ON.
Grounds	Test all the ECM terminals labeled ground or return.
Internal Regulators	The ECM has internal regulators to supply a regulated, constant voltage to certain circuits.
Self Diagnostic Section	With the key ON, engine OFF, ground the ALDL terminal B and observe the CEL. If the ECM issues a code 12, this usually indicates that the self-diagnostic subsystem is OK.
ECM Drivers	Most of the ECM drivers can be tested using the above mentioned procedure. When the B terminal is grounded, the drivers are ON.
Data Stream	If the serial transmitter works, the data stream will be activated whenever the B line is at 2.5 volts. On most EFI engines, the data stream can be activated in State 8 or 9.

ECM Circuit Troubleshooting (Cont.)

Always test these basics before automatically condemning an ECM. Most replacement ECMs are non-returnable once they are installed. Test all of the driven devices and their related circuitry for shorts to prevent damaging a replacement ECM. Manufacturers such as Borroughs sell tools which will measure the current draw through the circuit. (Borroughs Tool number BT-8405). Be careful when handling the ECM or PROM. Use a static grounding strap to prevent accidental damage from electro-static discharge. When installing or removing a PROM, CALPAC or MEMCAL, be careful not to bend or damage the pins or the socket.

Coolant Temperature Sensor

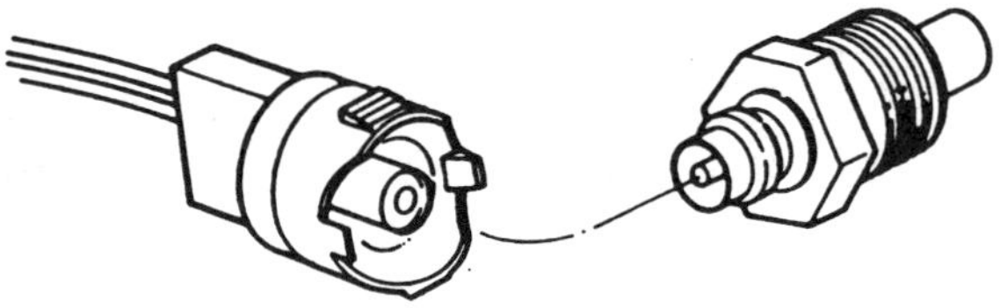

Courtesy of AC-Delco

What Is It?

The coolant temperature sensor is a negative co-efficient thermistor which is located somewhere in the engine water jacket. The thermistor is supplied a 5V signal by the ECM which measures the voltage drop across the circuit. The voltage normally "seen" by the ECM on the CTS signal line is approximately 4.0 volts at 70° F and approximately 1.0V at 212° F. The open circuit voltage reading would be full reference voltage and the shorted circuit voltage reading would be zero volts.

What's Its Job?

The ECM uses the input signal provided by the CTS to properly calculate A/F ratio, engine timing, idle speed commands, and also to control the TCC as well as the EGR, EFE, and canister purge solenoids. Obviously, this sensor is a very important sensor for overall ECM control.

Circuit Fault Codes:

Code 14 or Code 15

What To Watch For?

The CTS sensor has no moving parts. The housing that contains the thermistor can leak or become coated with cooling system sludge or corrosion. Contrary to popular public opinion, permanent antifreeze is not permanent. The earliest CTS sensors used Weatherpack connectors which were commonly guilty of poor connections. The later style CTS sensors use the Micropack terminals which have proven to be far more reliable.

Coolant Temperature Sensor Circuit

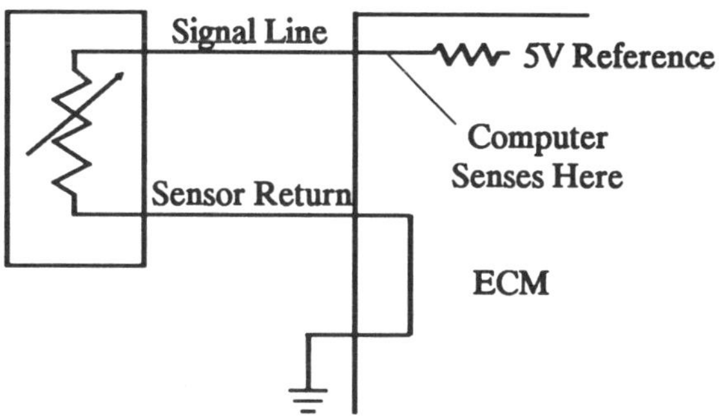

The ECM monitors the CTS signal line voltage anytime the ignition key is "ON". If the ECM detects a signal line voltage indicating a temperature of above 275° F, or below –35° F, the ECM will set a code. The possible causes of abnormally high or low signal line voltages are an open or shorted sensor, wiring or ECM internal circuits. It is also possible for the thermistor to become inaccurate and cause the signal line voltage to be neither high, nor low, yet be dropping the signal line voltage to an inaccurate level. The ECM has no ability to double-check the signal line voltage to guarantee the signal accuracy.

To test the CTS circuit, use a scan tool or a multi-meter. With the KEY ON, the voltage should be within a believable range. If the engine coolant is cold, the voltage will be around 4.0V. If the engine is warmed up to operating temperature, the signal line voltage should be near 1.0V. The scan tool may only read the circuit value as a temperature which is the converted signal as the ECM sees it. In many instances, the scan tool can display the CTS value in either degrees or volts.

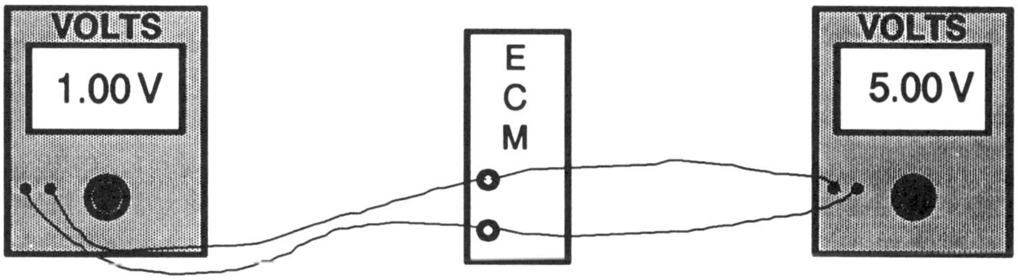

Warm Engine **Open Curcuit**

If the voltage is not in that range, or if a code has set, disconnect the coolant sensor connector. With the connector open, the signal line voltage should be full reference of 5.0V. If the reference voltage is present at the terminal of the connector which terminates the yellow wire, then the ECM is supplying the reference voltage and

Coolant Temperature Sensor Circuit (Cont.)

the yellow wire has continuity. Use a jumper wire to short across the two terminals of the sensor connector. If the black signal return wire has continuity to ground, then the signal line voltage will drop to 0.0V. If it does, then use the ohmmeter scale on the multi-meter to measure the resistance of the thermistor in the sensor. Compare the resistance reading with the Sensor Response Chart. If the resistance is infinite, the sensor is open. If the resistance is zero, then the sensor is shorted. If the resistance value is out of range, then the thermistor is out of calibration.If the resistance reading is within the response chart range, and the ECM is setting a code, yet the circuit tests proved that the wiring was OK, then the ECM can be assumed to be defective in that it is failing to recognize the proper CTS signal line voltage.

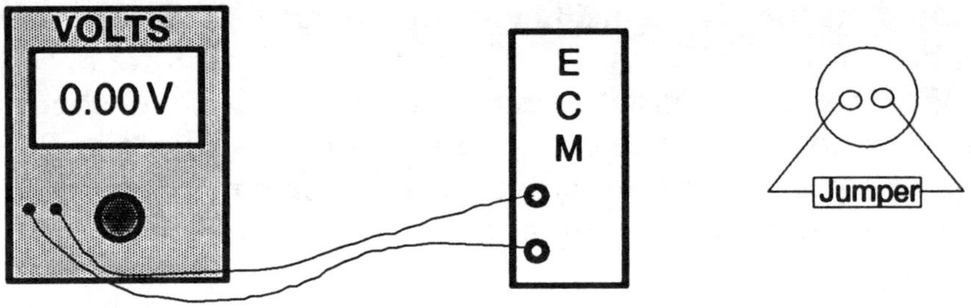

Shorted Circuit

In summary, if the CTS signal line voltage is correct at the ECM CTS signal terminal, then the circuit and sensor can be judged to be good. The voltage reading taken at that terminal is the sum of the whole circuit. When using the scan tool, watch out for the later model ECMs which have the capability to substitute a value when the circuit has been detected in default. If you only read the scan tool value, the ECM may be placing a fixed believable value for CTS in the data stream when a live reading on the signal terminal might reveal the actual circuit voltage.

Coolant Temperature Sensor Response Chart

Approximate Temperature to Resistance Values

F°	C°	Ohms
210	100	185
160	70	450
100	38	1,800
70	20	3,400
40	4	7,500
20	–7	13,500
0	–18	25,000
–40	–40	100,700

Manifold Absolute Pressure Sensor and BARO Sensor

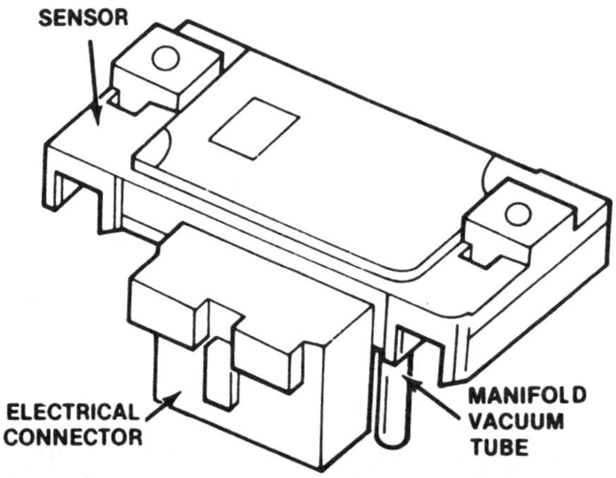

Courtesy of AC-Delco

What Are They?

The MAP sensors used on GM EFI engines are strain gauge, thick film pressure transducers which vary the voltage drop on the signal line from zero to five volts proportionate to the pressure applied to the sensor port. The ECM supplies a five volt reference through the sensor to the sensor return ciruit. The resistance internal to the sensor varies according to the pressure applied on the sensor port as manifold absolute pressure varies. The pressure applied is compared to the sealed pressure reference chamber so the sensor is an absolute pressure sensor. The MAP sensor hose should always be attached to a manifold port which has full manifold vacuum.

What Are Their Jobs?

The ECM uses the MAP signal to determine the engine load. It can also use the MAP signal line voltage to determine the barometric pressure by monitoring the voltage when the key is "ON" and the engine is "OFF". The ECM uses the MAP signal to determine the proper command calculations for fuel and timing control. On a GM EFI engine equipped with only a MAP sensor, the ECM is programmed to calculate fuel and timing commands based on the Speed Density control tables. On a GM engine which has a MAF sensor, the ECM uses the MAF signal to calculate the fuel commands based on the mass and flow of the incoming engine air. On a GM EFI engine which has both a MAP and a MAF, the ECM will probably use the MAF for fuel control and the MAP for timing control. The ECM can use the MAP as a back up for the MAF in the event of a MAF failure.

MAP and BARO Sensors

Cont.

Circuit Fault Codes:

Code 32, Code 33, Code 34, Code 63, or Code 64

Note: Be careful here, since these code numbers can also be something other than a MAP failure.

What To Watch For?

Most MAP sensor codes usually turn out to be a vacuum line or vacuum signal problem. The MAP sensor can be damaged by shock or by excessive pressures. Also, misrouted vacuum hoses can cause the MAP sensor signal to be inaccurate, yet not out of range far enough to set a code.

MAP and BARO Sensor Circuits

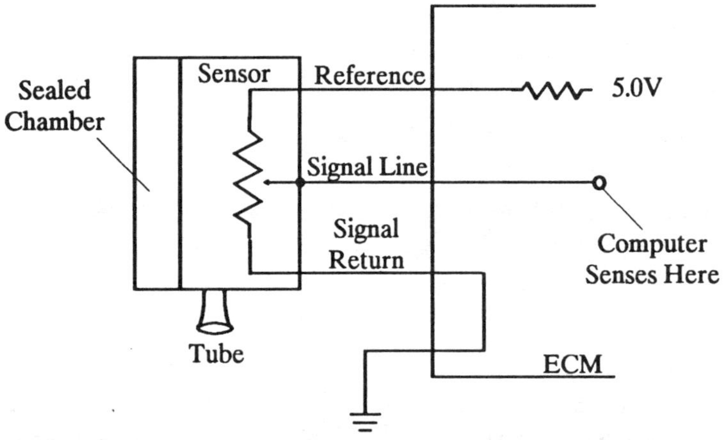

To test the MAP sensor and the related circuitry, use a multimeter or a scan tool. The scan tool may display the signal as a voltage or as a pressure reading in either hg or in kilopaschals. The voltage at the MAP signal line terminal will usually be approximately 4.6 to 4.8V with the key ON and the engine OFF.

If the voltage is not within that range, or if a fault code has been set, disconnect the sensor harness connector and measure the VREF between the gray wire and the black wire. It should be 5.0V.

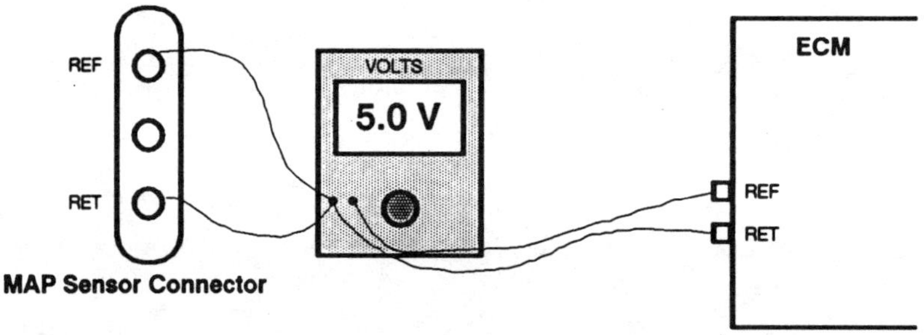

Use a jumper wire to connect the gray wire to the signal line wire. The reading at the ECM terminal should rise to full reference voltage if the signal line wire is not open or shorted.

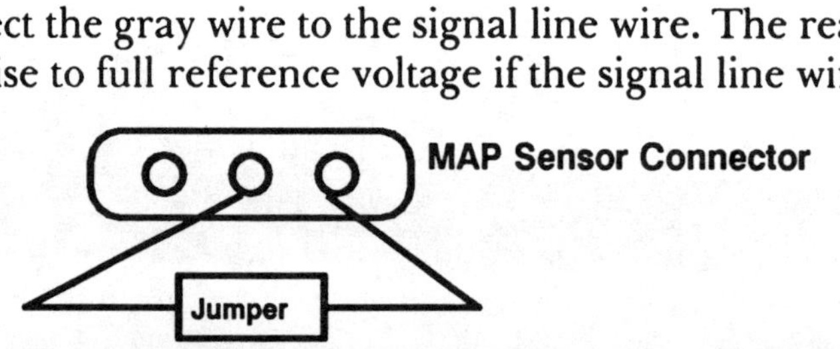

MAP and BARO Sensor Circuits (Cont.)

Test the sensor's accuracy by applying varying amounts of pressure to the sensor and comparing the resultant output voltage to the Sensor Response Chart.

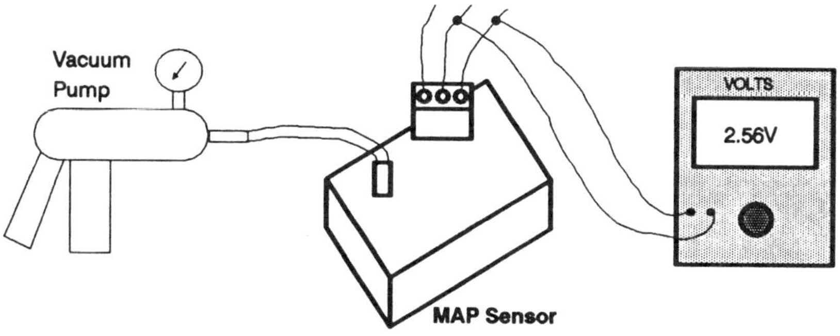

In summary, if the ECM is supplying VREF and is capable of "seeing" and responding to MAP signal changes, it is doing its job. If the MAP sensor voltages correspond to the Response Charts, then it is capable of doing its job. If the vacuum tubing is properly connected to the right port and the vacuum delivered is not restricted or leaking, then the vacuum supply is OK. If the jumper tests proved the circuit wiring to be not open or shorted, then it is OK.

MAP and BARO Sensor Response Chart

Red – 16006835
White – 16006832
Blue – 16006833
Green – 16017460

	Reference Power Supply Voltage		
Pressure in Hg	4.75V	5.00V	5.25V
3.0	4.654	4.900	5.147
2.5			
2.0	4.588	4.829	5.070
1.5			
1.2	4.506	4.741	4.977
0	4.330	4.556	4.782
–1.2	4.137	4.353	4.568
–2	3.995	4.203	4.411
–3	3.828	4.027	4.227
–4	3.649	3.839	4.029
–5	3.481	3.662	3.843
–7	3.129	3.292	3.454
–8	2.962	3.116	3.269
–9			
–10	2.616	2.752	2.887
–11			
–12	2.280	2.398	2.517
–13	2.104	2.212	2.321
–14			
–15	1.752	2.212	1.933
–16			
–17	1.420	1.843	1.566
–18	1.250	1.493	1.378
–19			
–20	.907	1.314	.998
–21			
–22	.570	.952	.626
–23	.385	.403	.422

VAC Sensor Response Chart

GM PN 16005-834
Gray or Black Insert

	Reference Power Supply Voltage		
Pressure in Hg	4.75V	5.00V	5.25V
3	0.032	0.028	0.023
2	0.073	0.174	0.174
1	0.352	0.362	0.372
0	0.529	0.548	0.567
1	0.696	0.723	0.750
2	0.940	0.903	0.939
3	1.036	1.080	1.130
4	1.098	1.251	1.303
5	1.388	1.450	1.514
8	1.904	1.994	2.086
10	2.234	2.345	2.456
12	2.608	2.722	2.843
15	3.094	3.253	3.415
20	3.966	4.156	4.353
21	4.135	4.235	4.555
22	4.317	4.537	4.755

PRESSURE RELATIONSHIP CHART

Pounds Per Square Inch Absolute	Pounds Per Square Inch Gauge and Inches of Mercury Gauge	Kilopascals Gauge and Centimeters of Mercury Vacuum	Kilopascals Absolute	Scanner Displays MAP Sensor In These Units Inches of Mercury Absolute
29.70 psia	15 psig	104.43 kPag	204.76 kPaa	60.45 in Hg
26.70 psia	12 psig	82.74 kPag	184.07 kPaa	54.35 in Hg
24.70 psia	10 psig	68.95 kPag	170.28 kPaa	50.28 in Hg
21.70 psia	7 psig	48.27 kPag	149.60 kPaa	44.17 in Hg
19.70 psia	5 psig	34.48 kPag	135.81 kPaa	40.10 in Hg
16.70 psia	2 psig	13.79 kPag	115.12 kPaa	33.99 in Hg
14.697 psia	0 psig/in Hg	0.00 kPag/cm Hg	101.33 kPaa	29.92 in Hg
13.71 psia	2 in Hg	5.08 cm Hg	94.53 kPaa	27.92 in Hg
12.24 psia	5 in Hg	12.70 cm Hg	84.39 kPaa	24.92 in Hg
11.26 psia	7 in Hg	17.78 cm Hg	77.63 kPaa	22.92 in Hg
9.78 psia	10 in Hg	25.40 cm Hg	67.43 kPaa	19.92 in Hg
8.80 psia	12 in Hg	30.48 cm Hg	60.68 kPaa	17.92 in Hg
7.33 psia	15 in Hg	38.10 cm Hg	50.54 kPaa	14.92 in Hg
5.85 psia	18 in Hg	45.72 cm Hg	40.34 kPaa	11.92 in Hg
4.87 psia	20 in Hg	50.80 cm Hg	33.58 kPaa	9.92 in Hg
3.89 psia	22 in Hg	55.88 cm Hg	26.82 kPaa	7.92 in Hg
2.42 psia	25 in Hg	63.50 cm Hg	16.68 kPaa	4.92 in Hg
0.00 psia	29.92 in Hg	75.996 cm Hg	0.00 kPaa	0.00 in Hg

Note: All Values are taken at sea level.

Oxygen Sensor

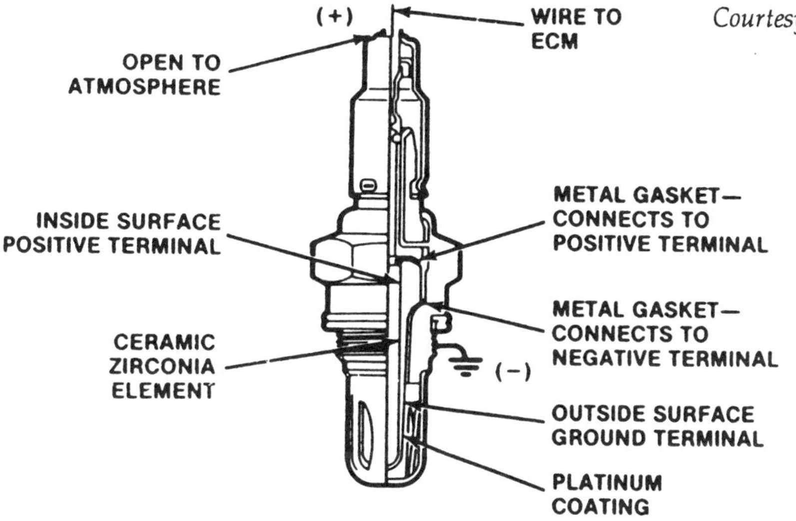

Courtesy of AC-Delco

What Is It?

The oxygen sensor is a voltage producing device which is located in the vehicle exhaust system. It is composed of Platinum electrodes which are separated by a Zirconium Dioxide cell. The sensor produces a low voltage (0.0 to 1.0V) signal which is proportionate to the amount of oxygen in the exhaust stream. The ECM supplies a reference voltage of approximately .45V on the signal line. The sensor has to be hot in order for it to produce a voltage. When the exhaust oxygen level is low, the sensor produces a high voltage. When the exhaust oxygen level is high, the sensor produces low voltage.

What's Its Job?

The oxygen sensor is used on the system so the system can control the fuel management system within the tight air/fuel ratio of Lambda or stochiometric. The main purpose of this tight fuel control is to protect the operation of the three-way, dual-bed catalyst system.

Circuit Fault Codes:

Code 13, Code 44 or Code 45

Oxygen Sensor

Cont.

What To Watch For?

Most oxygen sensors are killed by contamination from fuel or oil additives. Silicon sealers and the like are generally the culprits. The Zirconium cell is porous and if it becomes plugged up with foreign material, the cell will produce no voltage or an incorrect voltage.

Oxygen Sensor Circuit

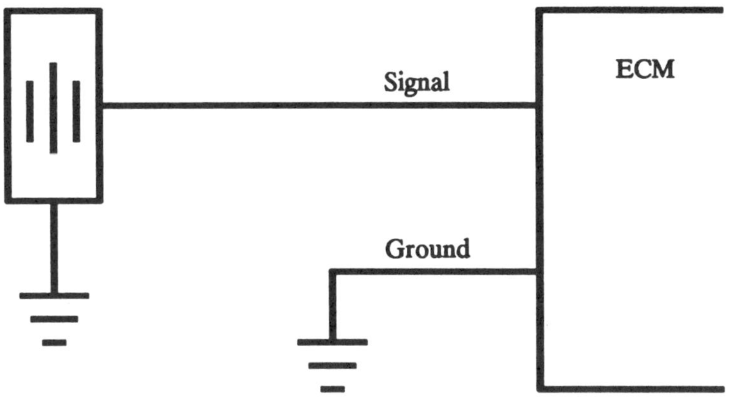

Test the oxygen sensor circuit with a voltmeter or a scan tool. The signal line voltage with the sensor disconnected should be approximately .45V with the key ON. With the key ON, engine RUNNING at normal operating temperature, the oxygen voltage should be switching back and forth across the .45V switch point.

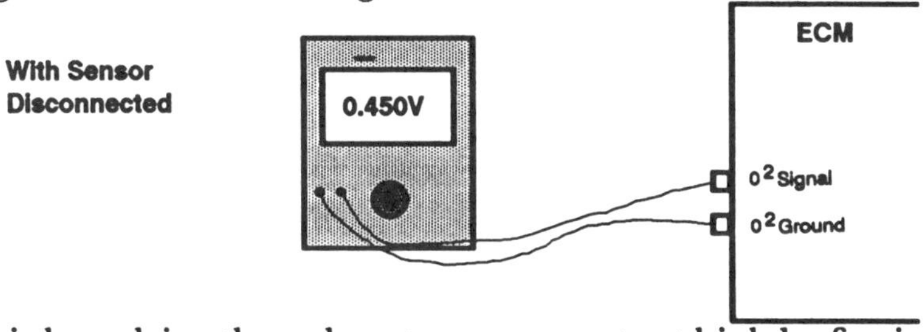

If the voltage is low, drive the exhaust oxygen content high by forcing the ECM to command rich. The oxygen voltage should rise to above .750V. If the voltage is high, drive the exhaust oxygen content low by creating a vacuum leak or force the ECM to command the system lean. With high exhaust oxygen content present in the system, the oxygen sensor should reduce the signal line voltage to below .10V. While observing the voltage produced, also notice the rate at which the oxygen sensor voltage will change. The oxygen sensor has been known to become "lazy". The voltage production should indicate that the sensor responds to oxygen changes rapidly.

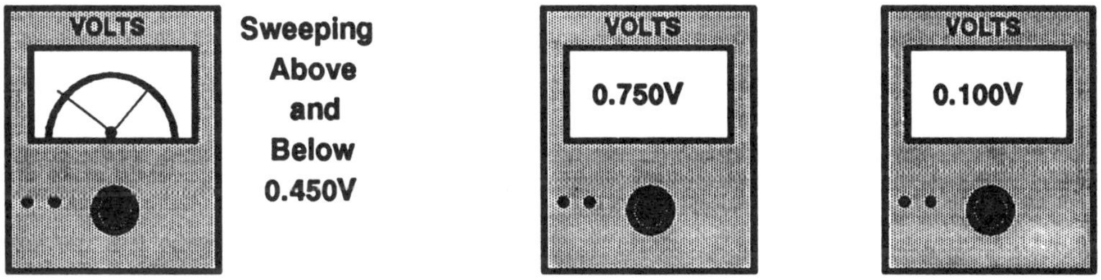

Always be on the lookout for abnormal sources of air such as the Air Management System or an exhaust leak. Also watch out for abnormal sources of fuel such as the canister system or contaminated engine oil.

Throttle Position Sensor

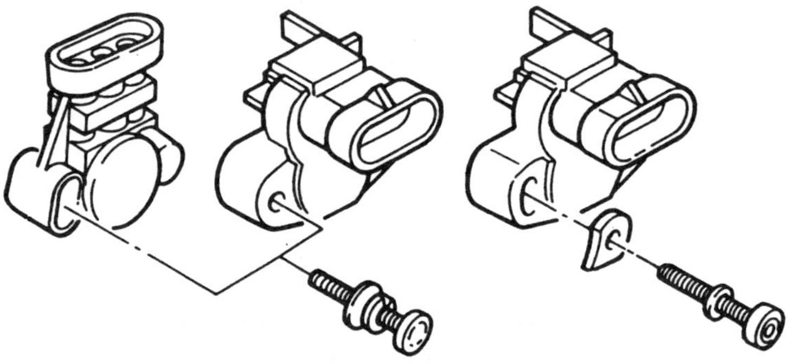

Courtesy of AC-Delco

What Is It?

The throttle position sensor or TPS, is a potentiometer which is attached to the throttle body assembly in a manner which allows movement of the throttle blade(s) to activate a swing arm which is attached to the sensor. The ECM supplies a 5.0V reference signal through the sensor to the sensor return circuit. The wiper arm of the potentiometer moves accross the resistance substrate dropping the signal line voltage proportionate to the throttle position. When the throttle is closed or at the idle position, the signal voltage will typically be at about .5-1.0V. When the throttle is open all the way (WOT) the signal line voltage will be above 4.0V.

What's Its Job?

The TPS signal is used by the ECM to properly calculate the commands for air/fuel ratio, timing, idle speed, TCC operation, AC Clutch, and emission systems control.

Circuit Fault Codes:

Code 21 , Code 22 or Code 23

What To Watch For?

Since the TPS sensor does contain some moving parts, wear and tear can occur internally. The resistance substrate can become worn or cracked. Some TPS sensors are adjustable and can be moved out of or into calibration.

Throttle Position Sensor Circuit

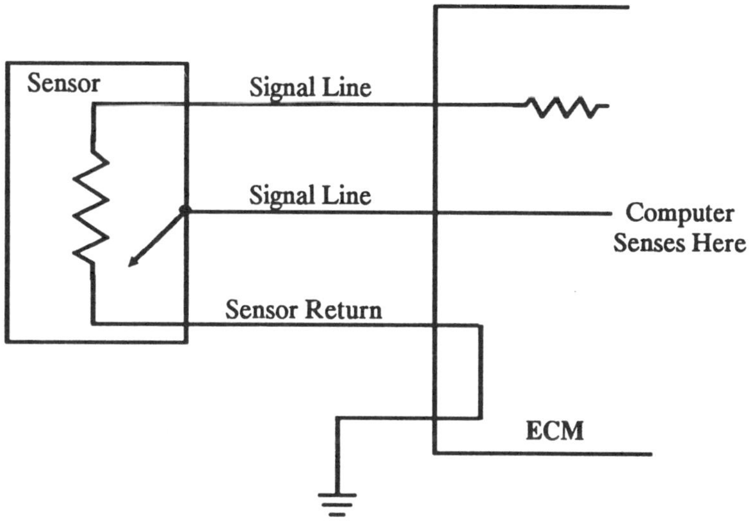

Circuit 1

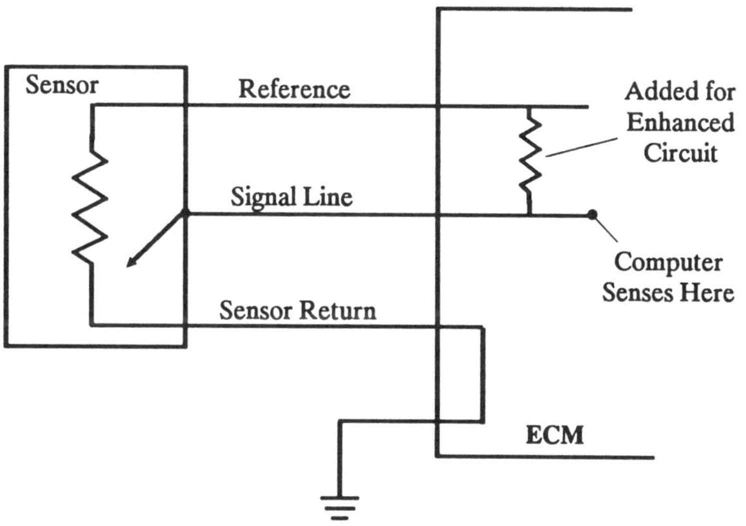

Circuit 2

The two circuit illustrations above represent the two possible TPS circuit diagrams. Under normal operation, the ECM "sees" the voltage proportionate to throttle position on the signal line. The difference between Circuit 1 and Circuit 2 is the extra resistor inside the ECM which is connected between the VREF line and the signal line. However, if the signal line is open, the voltage at the ECM TPS signal terminal will be 0.0V for Circuit 1 and 5.0V for Circuit 2.

To test the TPS circuit, use the scan tool or voltmeter to measure the TPS signal voltage with the KEY ON, ENGINE OFF. The voltage at idle should be within the

Throttle Position Sensor Circuit (Cont.)

factory specification. If it is not, then adjust the TPS if possible. If the voltage is within the spec, then slowly open the throttle to wide open throttle (WOT). The TPS voltage should exceed 4.0V. When the throttle is allowed to return to idle, the TPS voltage should return to the original idle voltage. If the TPS signal does meet these requirements, the TPS sensor and the circuits are probably OK. There can still be a cracked or worn spot on the internal resistance substrate which can only be detected using an analog ohmmeter, or a scope.

To test for a cracked or worn spot, test between the signal terminal and the return of the disconnected sensor with an analog ohmmeter. Sweep the sensor from the resting position to WOT. The ohmmeter needle should sweep steadily across the scale. If the sensor wiper crosses a problem area, the needle will jerk or perhaps even jump to infinite. Sweep the sensor several times to make sure.

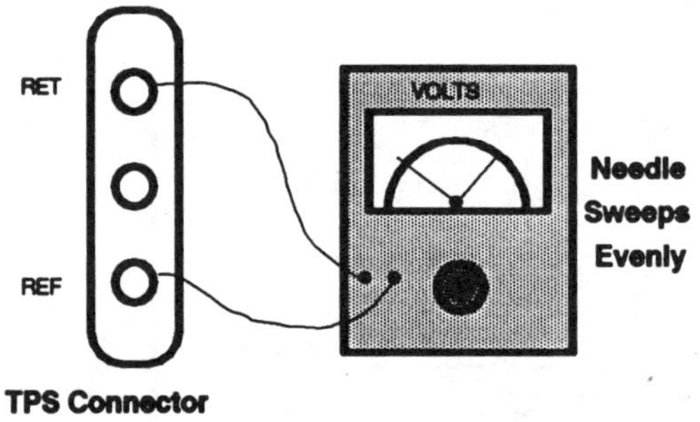

If the voltage on the TPS signal line was out of range, then test the VREF voltage at the disconnected sensor harness. If the voltage is not present, then the ECM is doing its job. If the VREF voltage is not present at either test point and the ECM is connected to a good power and ground, then the internal regulator is bad.

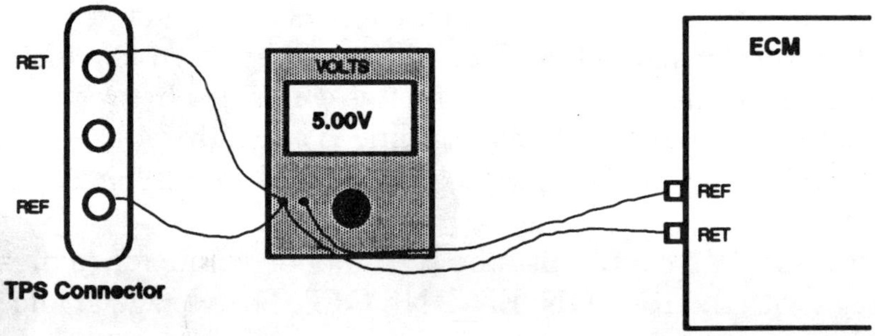

Throttle Position Sensor Circuit (Cont.)

If the 5.0V is available, then prove the circuits by jumpering the signal line to the return line with the KEY ON. If 5.0V was present and the voltage goes to zero, then the sensor return line is OK. If 5.0V was not present on the signal line, then jumper the signal line to the VREF line. If the signal line voltage goes to 5.0V, then the signal line is OK.

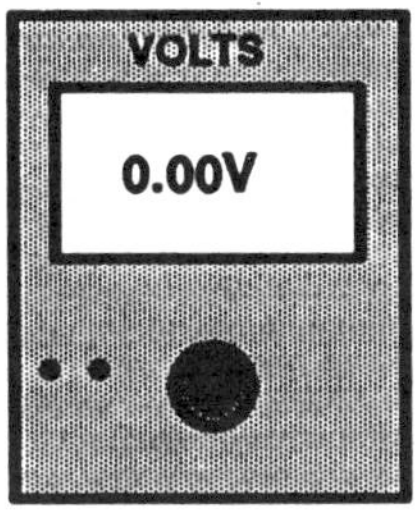

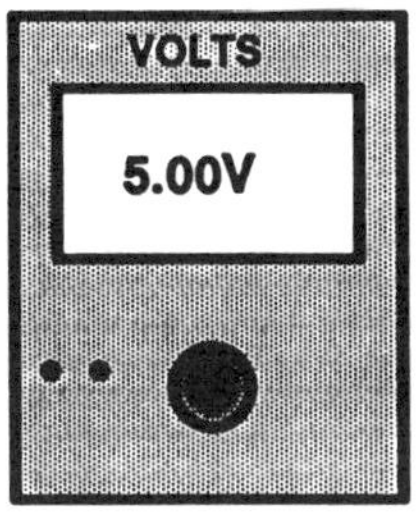

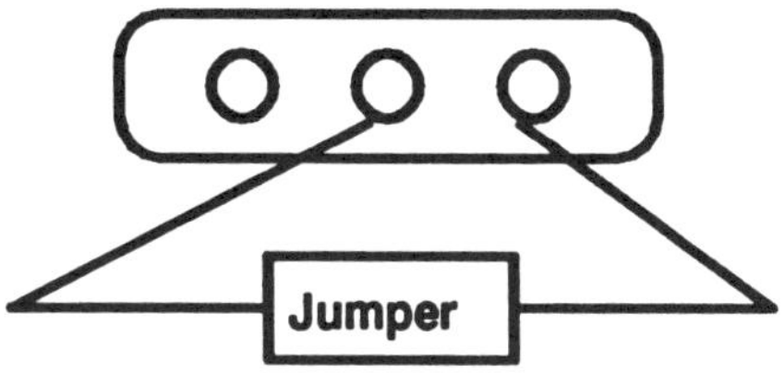

In summary, the ECM's job is to provide the VREF voltage. The sensor's job is to proportionately drop that voltage across the signal line depending on throttle position. Testing the signal voltage at the ECM terminal will shorten the test steps if the reading is good. The scan tool may seem to be delayed in its reading updates. This is due to the serial data baud rate, not the scan tool.

Mass Airflow Sensor

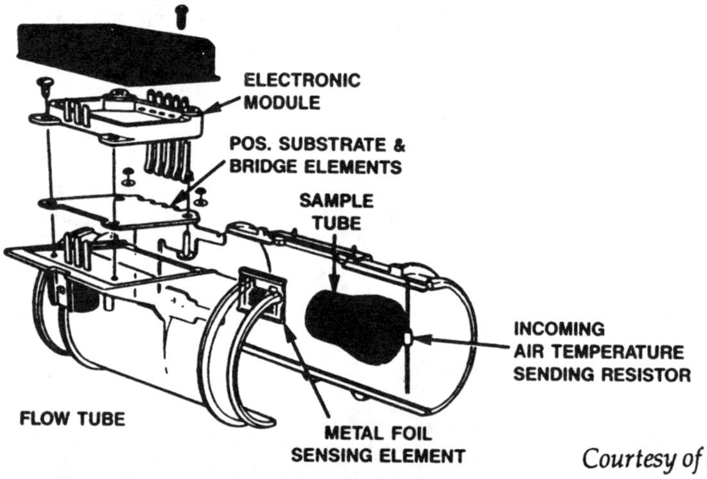

Courtesy of AC-Delco

What Is It?

The Mass Air Flow sensor, or MAF, used on GM EFI engines will be one of four types. The possible types are:

Delco 3-wire
Delco 5-wire
Bosch Hot Wire
Hitachi

The MAF sensors are used on a variety of engine applications. The Delco sensors and the Hitachi sensors are frequency generators. The Bosch sensor generates an analog voltage. The electronics contained within each sensor are quite sophisticated. The basic functioning principle is the measurement of the electrical power required to maintain the sensing element at a specific temperature. The Bosch sensor has extra terminals dedicated to the operation of a clean-off operation performed by the burn-off relay. The frequency range for the frequency generators is 30-150 hz. The analog signal for the Bosch is 0.0V to 5.0V.

What's Its Job?

The MAF signal informs the ECM of the airflow or load on the engine which the ECM uses to calculate injector control pulse width and engine timing requirements.

Circuit Fault Codes:

Code 33, Code 34 or Code 36

Mass Airflow Sensor

Cont.

What To Watch For?

The Delco MAF sensors have a notorious reputation for premature failure. The principle cause of failure on the Delco sensors has been poor solder connections within the sensor housing. When the solder joints work loose, any movement of the sensor can cause malfunction or intermittent function. Tapping on the sensor housing lightly with a tool will usually reveal this poor connection condition. The second most common failure has occurred because the sensor electronics become coated with foreign substances. The problems with Delco MAF sensors have been so prevalent that the Chevrolet-Pontiac-Canada Group of GM has all but quit using MAF sensors. The Hitachi MAF sensors have proven to be quite reliable. The Bosch Hot Wire sensor has also been very reliable and the limited application (Tuned Port FI only) has contributed to the limited number of field failures of this sensor design. The sensor does not always set a fault code when any of these failures occur since the ECM is programmed to accept a very wide range of operating signals from the MAF sensor.

Mass Air Flow Sensor Circuits

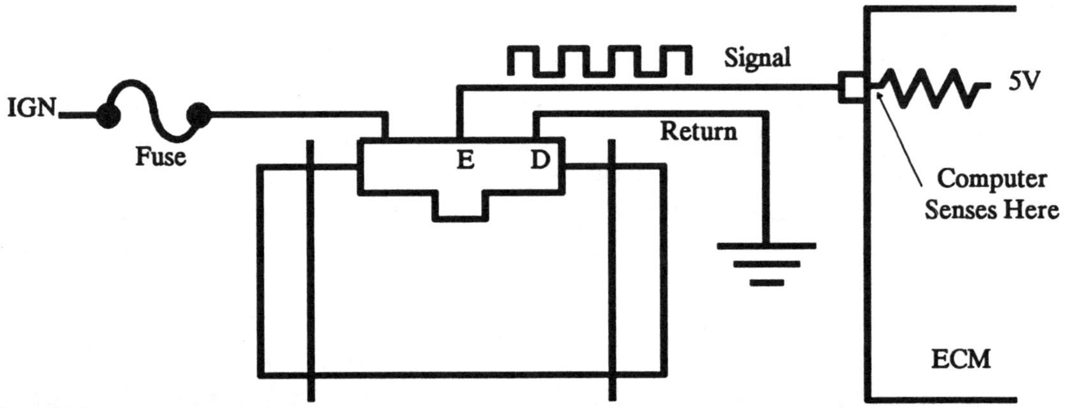

Circuit 1

Three-wire MAF without relay
Variable Frequency Signal

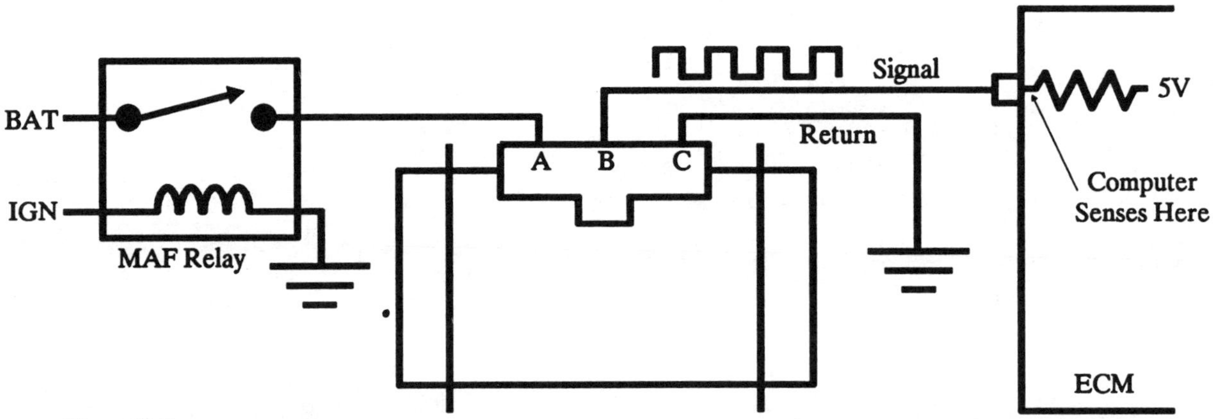

Circuit 2

Three-wire MAF with relay
Variable Frequency Signal

Mass Air Flow Sensor Circuits (Cont.)

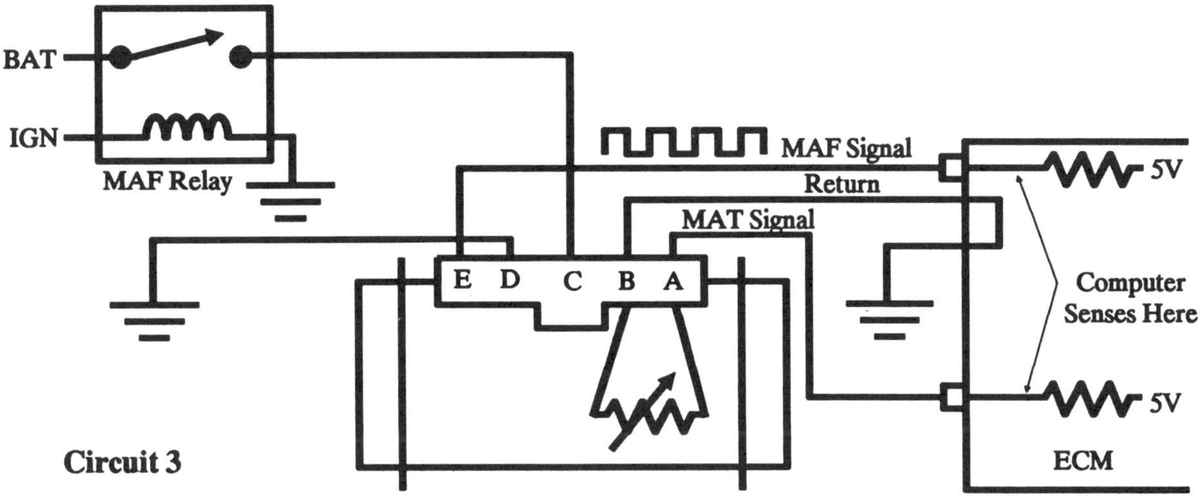

Circuit 3

Five-wire MAF
Variable Frequency Signal

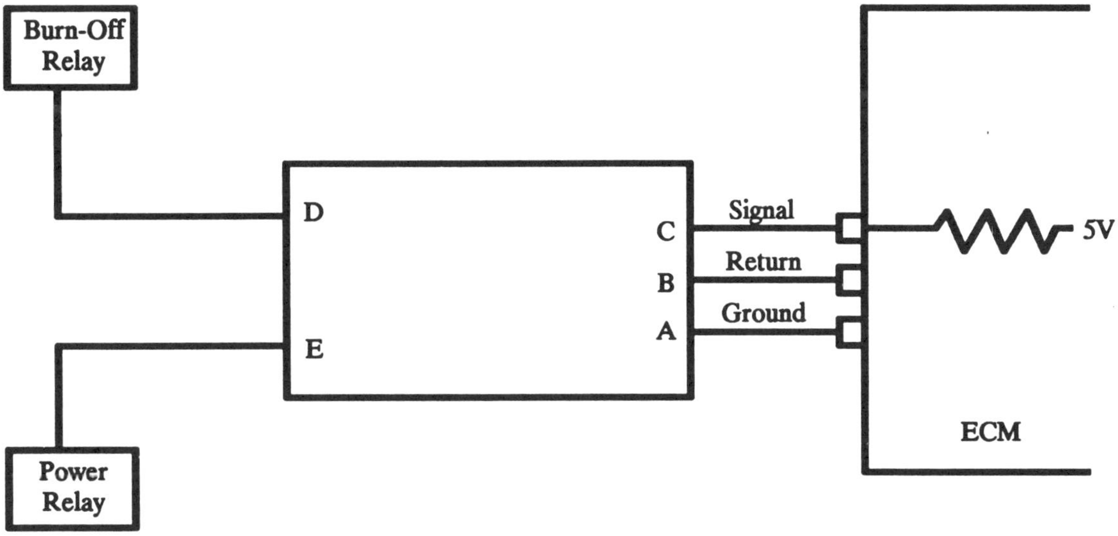

Circuit 4

Bosch "Hot" Wire
Power Relay & Burn-Off Relay
Variable Analog Voltage Signal

Mass Air Flow Sensor Circuits (Cont.)

There are four possible circuits used by GM for the MAF sensor. Circuit 1 is for the 3-wire MAF with no MAF relay. Circuit 2 is for the 3-wire MAF with a relay. Circuit 3 is for the 5-wire MAF. Circuit 4 is for the Bosch MAF with a burn-off relay.

To test the MAF sensor signal, the technician can use the scan tool to observe the signal through the data stream. The scan tool will display the signal as a converted reading of grams of air flow per second. This reading will typically be approximately 3-5 grams per second at idle and will ascend linearly as the engine RPM is increased. By graphing the MAF signal from idle RPM up through the full operating RPM range of the engine, the technician can observe the MAF signal trends. Keep in mind the fact that in the stall test, the engine will not draw as much air as in actual road load conditions. The main thing to watch for is a radical change in the MAF signal trend. Also keep in mind the fact that one gram of air is a lot of air. This means that the grams per second reading displayed through the data stream is a fairly low resolution reading. If the technician will use a frequency meter to measure the MAF signal changes in actual frequency, the resolution of the measurement becomes much greater.

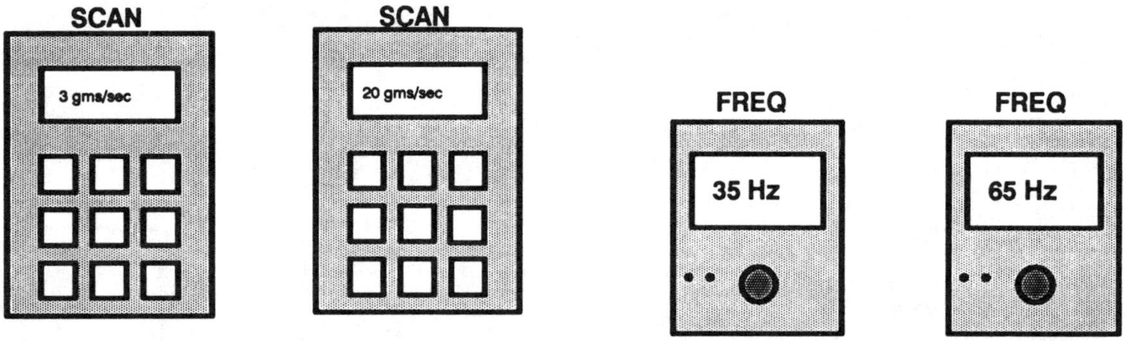

The MAF Response Chart can be used as a suggested sensor tracking method to more closely observe the MAF signal ascension. The MAF signal can also be displayed and observed on an O-scope to verify that the signal generated is "clean", or free from interference. The signal signature should look something like this:

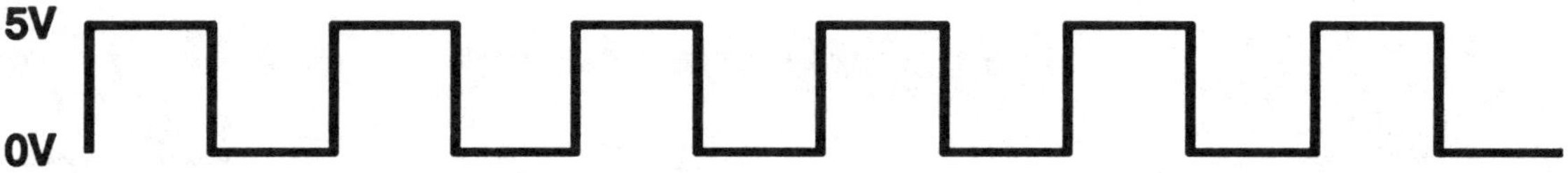

The MAF signal can also be verified by using a Logic Probe. The digital pulse produced by the MAF sensor should trigger all of the LEDs. If the MAF sensor is not producing a signal at all, test the power and ground circuits to the MAF. If relay

Mass Air Flow Sensor Circuits (Cont.)

equipped, the relay is responsible for supplying the power. The ECM is responsible for placing the 5 volts on the signal line. The electronics inside the MAF sensor are responsible for toggling the 5 volts to ground at a variable frequency corresponding to air flow.

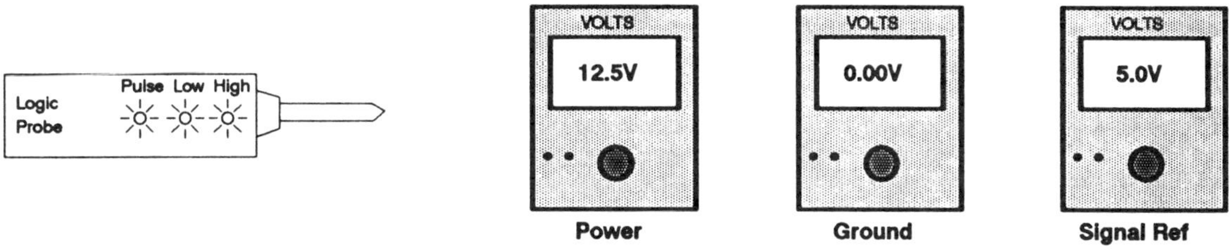

The scan tool will display two different readings on some vehicles. One reading will be called Mass Air Flow. The other reading will just be called Air Flow. The Mass Air Flow reading should be the same as the Air Flow reading, unless a Code 33 or Code 34 has been set. If a code has been set, the Air Flow reading represents the default value being used by the ECM as a Mass Air Flow substitute value.

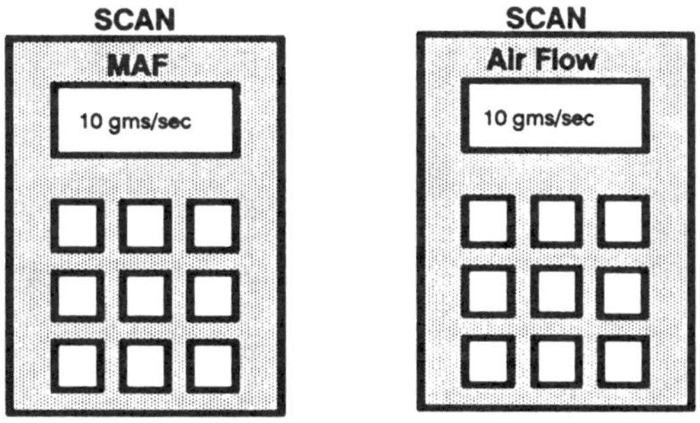

In summary, the MAF sensors used by GM are varied in source and type. The converted signal seen in the data stream by the scan tool is sometimes confusing to the technician until the operating philosophy is understood. Voltage, frequency and grams per second are all inter-related readings. Kent Moore even offers a MAF sensor tester which will sometimes help catch a bad sensor. This unit works on the 3-wire and 5-wire Delco units only.

Airflow Versus RPM Chart

Typical MAF Sensor

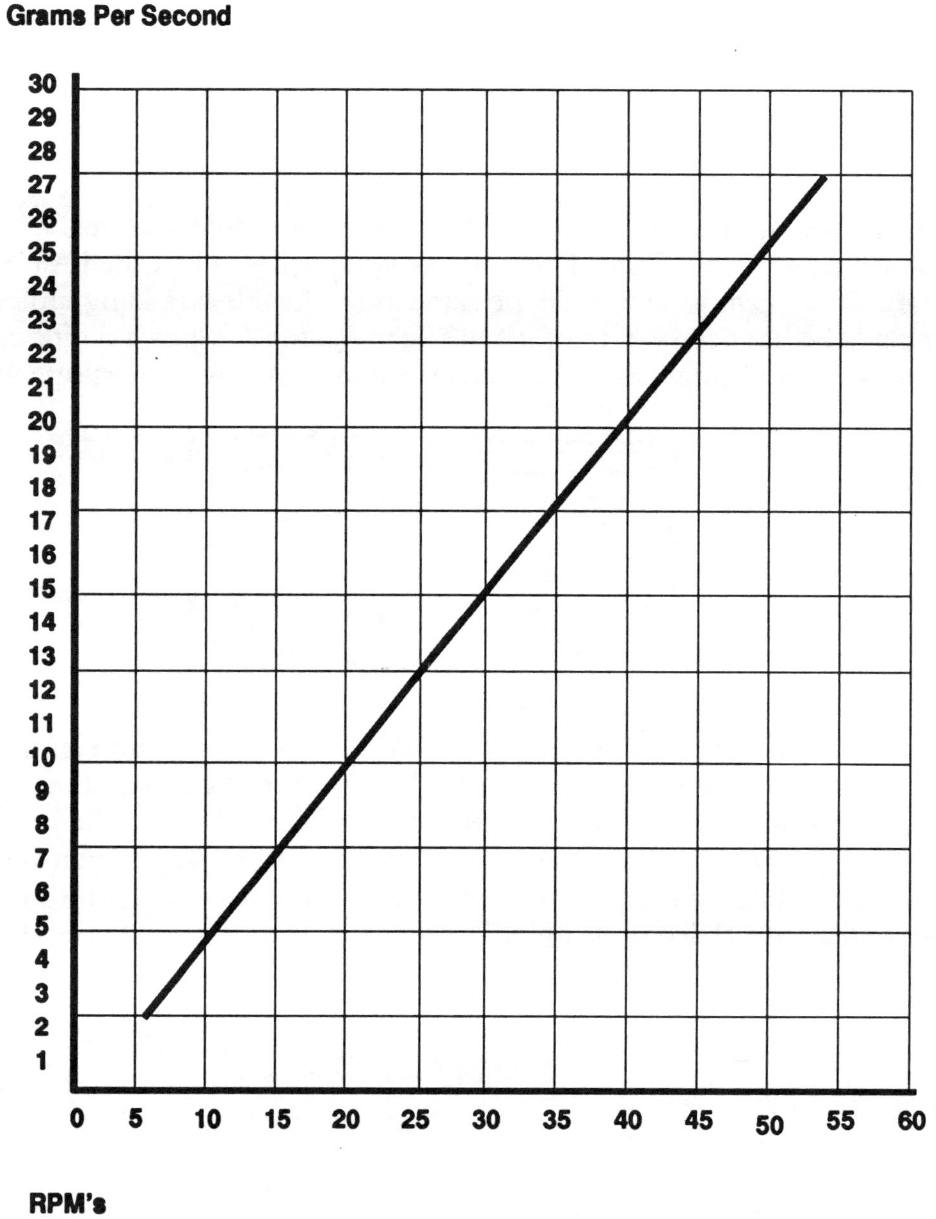

Note:
Should be an even linear slope, with no dips or blips.

MAT or Air Temperature Sensor

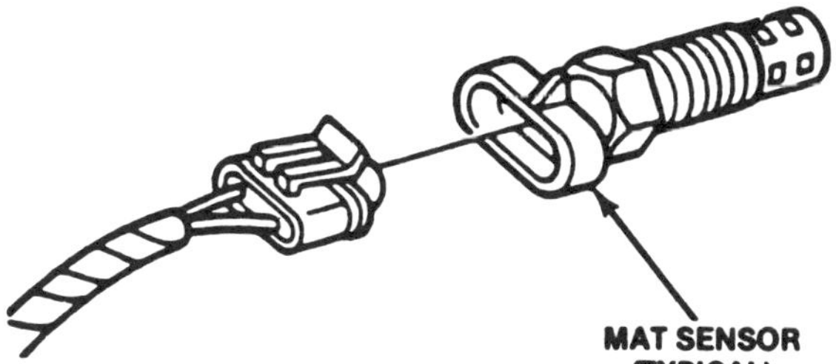

Courtesy of AC-Delco

What Is It?

The Manifold Air Temperature Sensor or MAT and the Air Temperature Sensor or ATS are negative coefficient thermistors located in the intake manifold or the intake air stream. Some will even be threaded into the throttle body assembly. The 5-wire Delco MAF has an air temperature sensor located within the Mass Air Flow Sensor housing. The ECM supplies a five volt signal on the sensor signal line and the thermistor drops the voltage across the circuit. If the circuit was open, the signal line voltage would be 5 volts. If the circuit was shorted, the signal line voltage would be zero volts.

What's Its Job?

The ECM uses the MAT or ATS sensor signal to modify the fuel and timing commands. This signal can also serve as a backup signal in the event of the loss of the Coolant Temperature signal.

Circuit fault Codes:

Code 23 or Code 25

What To Watch For?

Poor connections, pinched or broken wires. On some engines, the air sensor can become coated with foreign substances which affect the accuracy of the sensor.

MAT or ATS Circuit

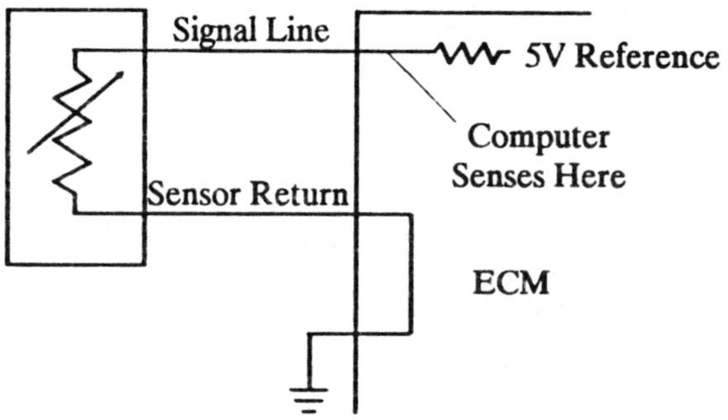

To test the MAT or ATS circuit, use the scan tool or a multi-meter. The voltage which should be on the signal will be between 1.0V and 4.0V, depending on the air temperature. If the voltage is abnormally high or low, disconnect the sensor harness connector and measure the reference voltage on the signal line. It should be 5.0V. If the VREF is not present, measure at the ECM terminal. If the ECM is supplying the voltage, then it's doing its job. If not, and the ECM has good power and ground, then the ECM internal regulator for this circuit is bad.

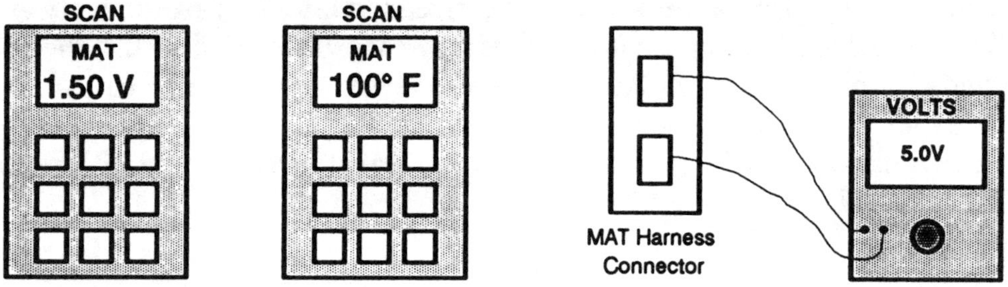

If the 5.0V are present, then prove the circuit by jumpering the signal line to the return line. If the return line is OK, then the voltage will drop to zero volts.

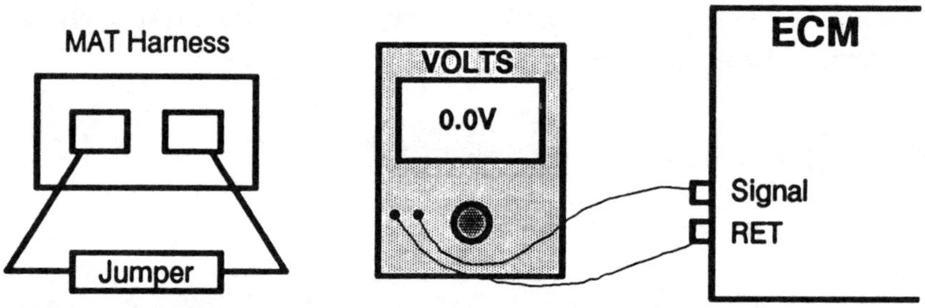

If the signal line voltage appears to be out of range, prove the thermistor by measuring the resistance and comparing the reading with the Sensor Response Chart. You may choose to use a pyrometer to verify or establish the ambient temperature.

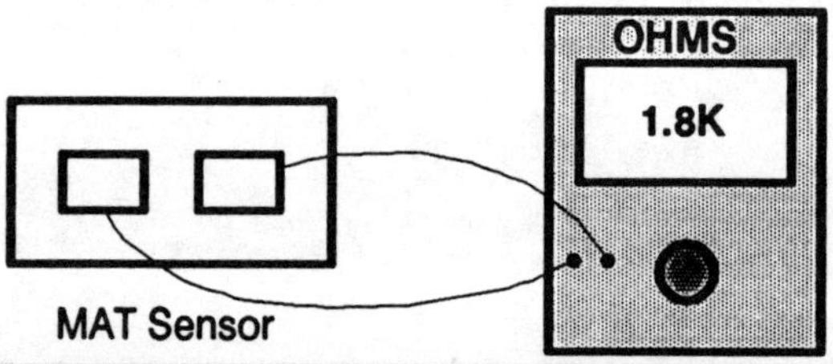

MAT Sensor Response Chart

Approximate Temperature to Resistance Values

°F	°C	Ohms
210	100	185
160	70	450
100	38	1,800
70	20	3,400
40	4	7,500
20	–7	13,500
0	–18	25,000

Detonation or Knock Sensor

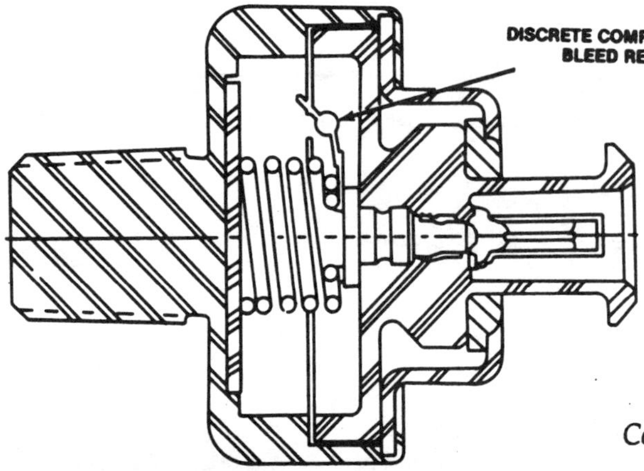

Courtesy of AC-Delco

What Is It?

The Detonation or Knock sensors used on GM EFI engines are piezoelectric sensors which work like a microphone. Vibrations or sound waves produced by the engine are converted to an AC voltage which is inputted to the ESC controller on some models and directly into the ECM on some others. The ESC controller or ECM puts a small reference voltage on the signal line which the AC sinusoidal waveform acts to overcome or modify.

What's Its Job?

The detonation sensor is only used on engines which have the timing retard system called Electronic Spark Control or ESC. The sensor will develop a voltage signal when it detects any vibration. The conroller that it feeds that signal to is designed to determine which vibrations are caused by engine detonation. The timing retard commands necessary to protect the engine from damage are then issued.

Circuit Fault Codes:

Code 43

What To Watch For?

Since a code 43 does not necessarily mean that the detonation sensor is bad, the technician must fully understand the conditions which can cause a code 43 to set. In Section II, there is a more complete explanation of what conditions set the fault code. The detonation sensor is actually blamed for a lot of code 43 problems or engine timing retard problems which it does not cause. The actual failure rate of this sensor is very low.

Detonation Sensor Circuits

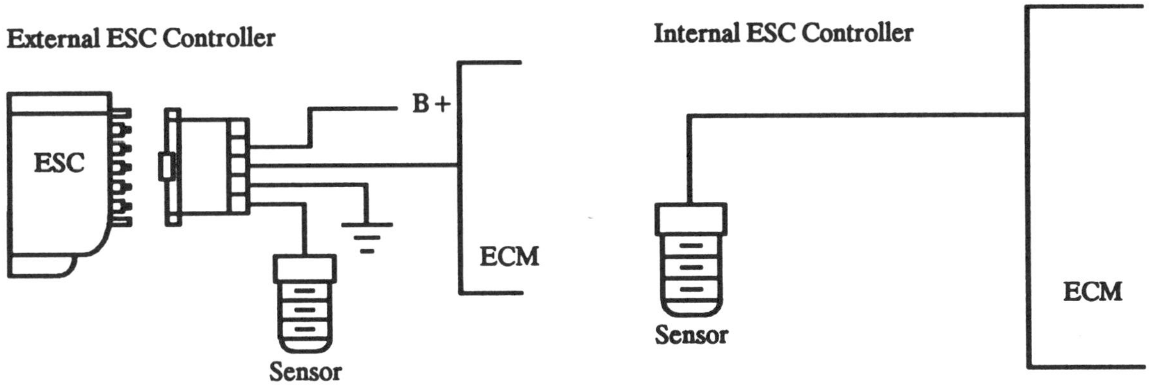

To test the Detonation Sensor circuit, the technician can take one of two paths. The first path would be to test the detonation sensor directly. This can be done with an AC voltmeter or with an O-scope. The detonation sensor produces a small amount of AC voltage when vibrations are sensed. Tapping lightly at or near the sensor with either instrument attached to the sensor terminal should verify whether or not the sensor is capable of generating the AC voltage. The voltage generated is less than 2.0V AC. The signal signature will look something like this:

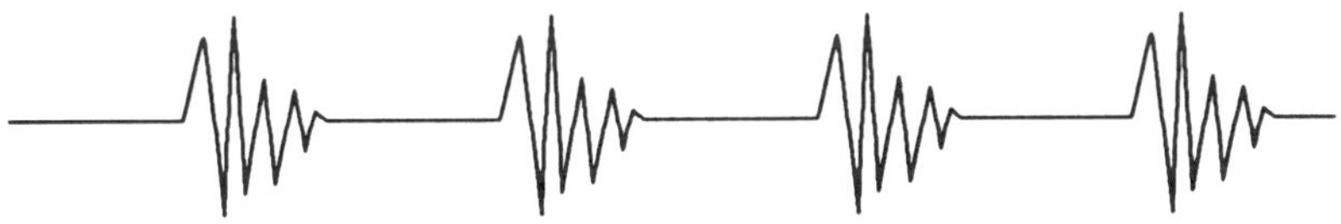

The other path of verification will involve the whole sensor input and control response system. For instance, with an advance style timing light hooked up, tapping at or near the sensor should produce an actual engine timing retard. If the timing retards, then this would verify that the sensor detected the vibrations, the ESC controller or ECM detected the sensor signal, and the resultant timing retard change occurred.

Another method of this type of verification would be to road test the vehicle with the ALDL "A-B" terminals jumpered together. This is called the Field Service Mode of operation. When those two terminals are jumpered, the ECM will artificially advance the engine timing. Upon hard acceleration, the engine should detonate. If the sensor is working, the ECM will detect the detonation signal and this should cause the ECM to attempt to retard the timing, which it can't, because the Field Service Mode of operation won't allow it to. The ECM should signal that the detonation is present by flashing the Check Engine or Service Engine Soon lamp at the rate of five times per second for three seconds. If the lamp flashes at this rate, this would verify that the ECM is recognizing and acknowledging the detonation signal.

Detonation Sensor Circuits (Cont.)

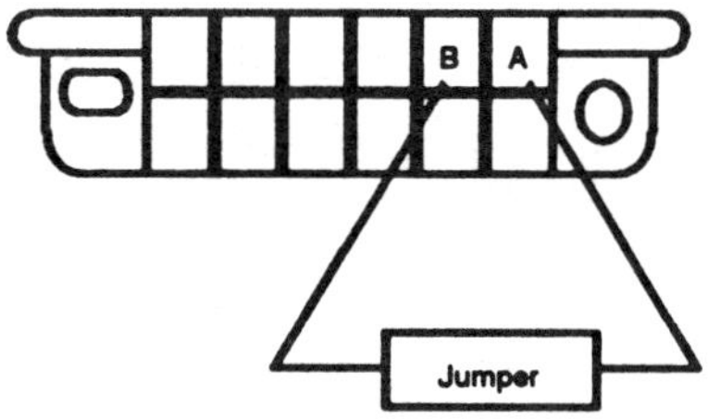

Five Flashes Per Second For Three Seconds

Another method of verification is to use the scan tool. On some models, when detonation is sensed, the ECM will retard the spark command signal on the EST line. Some ECMs display this in the data stream as the number of degrees of spark retard which is being commanded. The only trick here is you have to make the engine detonate in order to get the ECM to "see" a detonation signal and to issue the retard command. A test drive with hard acceleration should do it.

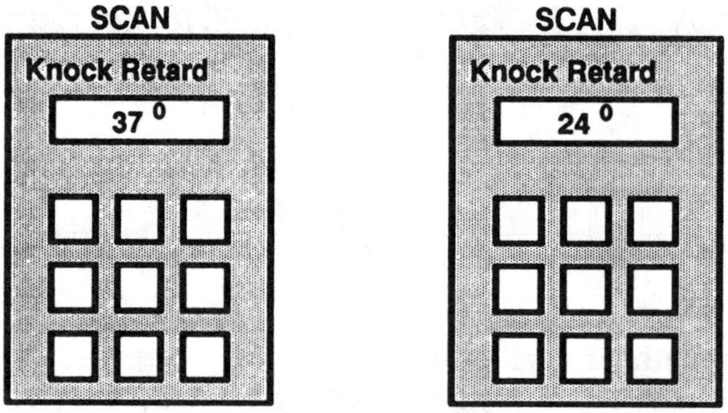

Vehicle Speed Sensor

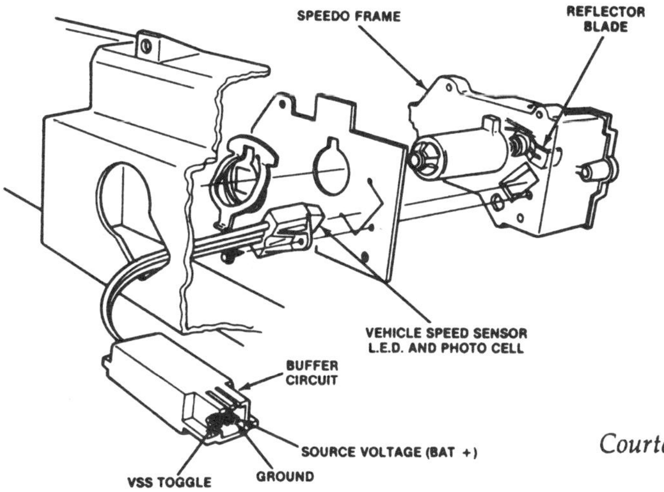

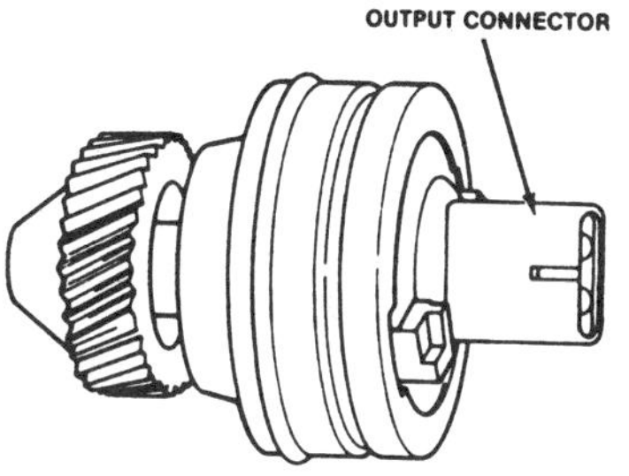

Courtesy of AC-Delco

What Is It?

The Vehicle Speed Sensor or VSS can be one of several different devices on GM EFI vehicles. The earliest form of the VSS was a photo diode located behind the speedometer in the instrument panel. An LED emitted light which was reflected by a rotating shiny disc attached to the back of the speedometer. A dark, non-reflective strip caused the light reflection to be interrupted with each revolution. The photo diode received the alternately interrupted light reflections and generated a DC voltage pulse which the ECM could "count" and determine the vehicle speed. The introduction of digital speedometers called for sensors of a different design since there would be no rotating disc in the instrument panel. The vehicles using digital speedometers use a wire wound voltage generator which is usually located in the transmission or transaxle housing.

What's Its Job?

The principle job of the VSS is to inform the ECM of vehicle speed so the ECM can control the operation of the TCC or Torque Converter Clutch. The later model ECMs also control the cruise control transducer based on this VSS signal. The VSS signal can also be a factor used by the ECM to determine when the vehicle is actually moving for IAC zero-referencing as well as factoring in some secondary input information for some of the emission control decisions.

Circuit Fault Codes:

Code 24

Vehicle Speed Sensor

Cont.

What To Watch For?

The VSS has proven to be a very reliable sensor. The buffers or amplifiers used to refine the VSS signals are more prone to failure than the actual sensors themselves. Many code 24 problems result from disconnected or broken speedometer cables. The PM generator type of VSS can become damaged or subject to poor connections due to the fact that the sensor is exposed to road debris since it is mounted in the transaxle.

Vehicle Speed Sensor Circuit

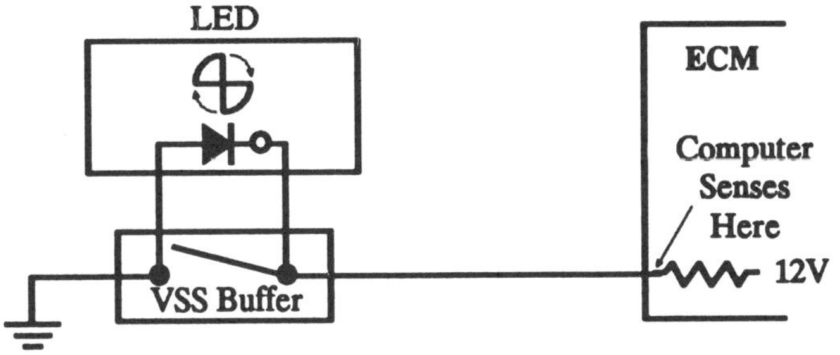

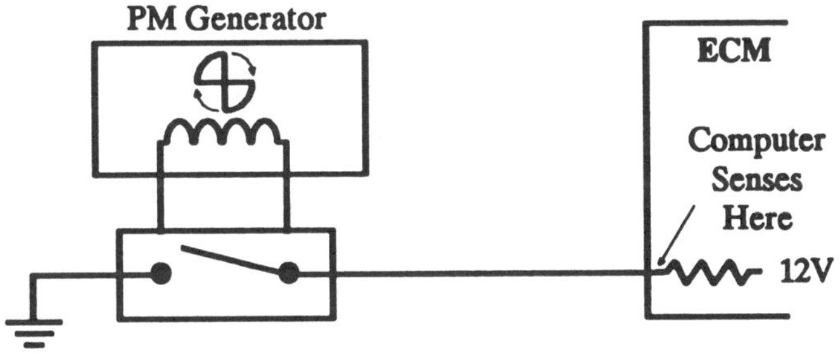

To test the VSS output, the technician can choose either a voltmeter, scan tool or O-scope. The scan tool will display the converted VSS signal input in MPH. The MPH reading displayed on the scan tool should match the actual speedometer reading. Watch out for the fact that the scan tool can display the speed in either MPH or KPH. If the scan tool displays the reading in either form, the VSS is working and the signal is being properly buffered and the toggling of the ECM circuits is occurring.

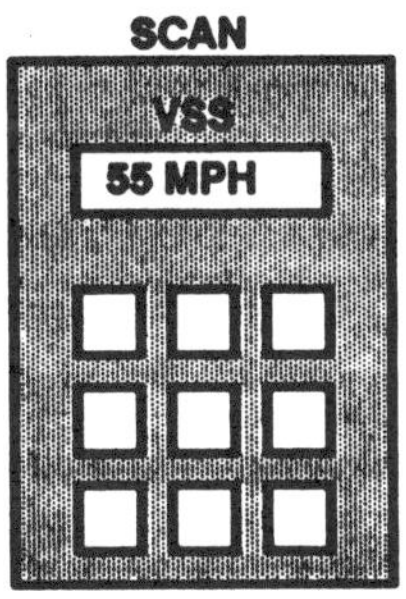

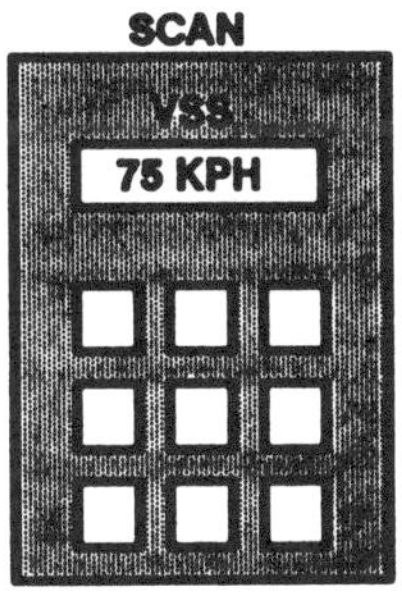

To test the VSS output directly, determine whether the VSS is the LED type or the PM generator type. The PM generator type generates an AC signal from 5.0V peak-to-peak to 8.0V peak-to-peak. Use an AC voltmeter or an O-scope to measure the output signal. The signal signature will look something like this:

Vehicle Speed Sensor Circuit (Cont.)

If the sensor is the LED type, the photo diode has to be powered by a DC pulse. The amplitude will be from 8.0V to 12.0V, depending on power source is used. The signal signature will look something like this:

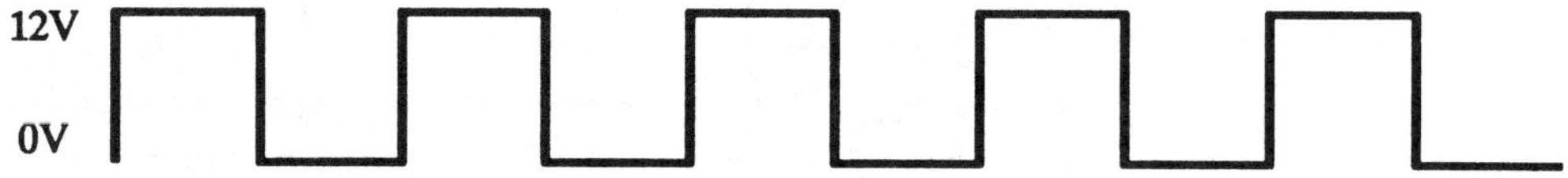

The actual ECM received signal generated by the VSS will be a buffered signal. The ECM supplies a 12.0V signal on the VSS line. The buffer circuit toggles the 12.0V signal to ground. The signal pulse will then measure an average of approximately 6.0V since the signal is a square wave. Some ECMs are designed to receive the PM generated AC signal directly into the ECM input terminals. The buffering can be done internally within the ECM since the processor cannot process an AC signal. Carefully examine the actual circuitry used to clarify what parts and circuits will be used externally on the ECM of the vehicle being examined or tested. Don't forget that the signal generated by the VSS can be shared by the actual vehicle speedometer, the ECM and in some cases the cruise control. Sounds confusing, but it's not. The VSS signal will be either a simple DC pulse or an AC sine wave.

Switch Signal Inputs

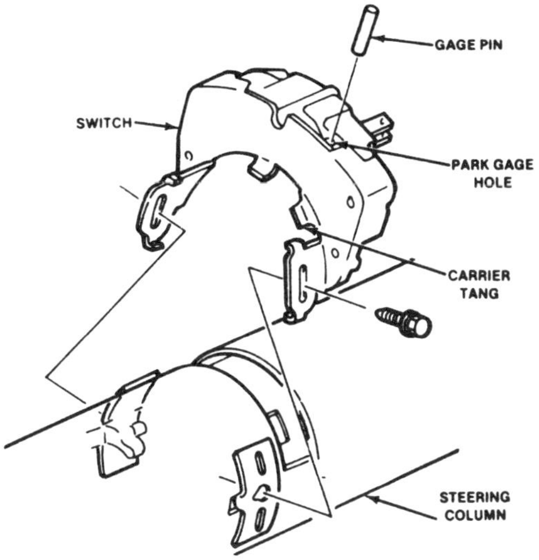

Courtesy of AC-Delco

What Are They?

The ECM configurations can have inputs which are simple switch signals. These signals come from a variety of circuits such as the Park/Neutral Switch, the Power Steering Load Switch, the Transmission Gear Switches (2nd, 3rd, and 4th), the AC Request or Control Switch, the EGR Vacuum Switch or the Cruise Control Status Switches (ON/OFF, Set, Resume, etc.).

What Are Their Jobs?

These switch signal inputs have a variety of ECM input assignments. The status of these input signals aid the ECM in determining the proper control responses for fuel control, timing management, TCC operation, IAC operation, AC operation and Cruise Control operation.

Circuit Fault Codes:

Code 27, 28, 29, 32, 35, 38, 46, 67, 68 and 69.

What To Watch For?

Poor connections, pinched or broken wires and dirty switch contacts will in most cases be the cause of these switch malfunctions.

Switch Signal Input Circuits

Testing switch signal input circuits can usually be accomplished using the scan tool. Some scanners display the status of many of the switch signal inputs with LEDs on the control panel of the scan tool. Some of the switch signal inputs can only be observed by selecting a scan tool mode for a particular switch signal status. A simple "ON" or "OFF" word or light will appear.

Examining the individual switch input circuits will reveal some variations from application to application. For instance, the Park/Neutral Switch circuit illustrated below:

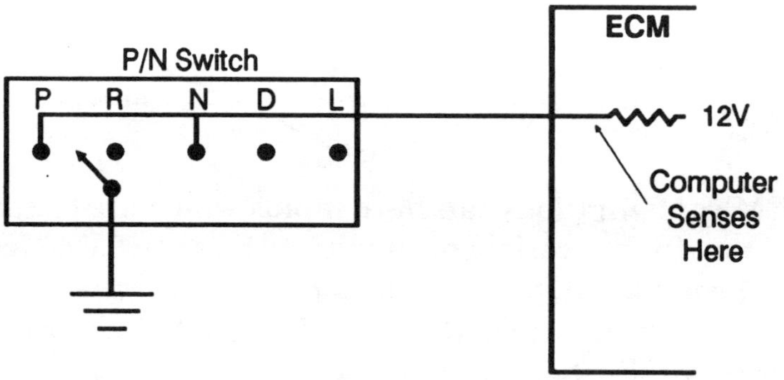

On this circuit, the ECM supplies a 12.0V, current limited signal on the signal line. When the gearshift selector is moved through its operating range, the switch will short the signal to ground whenever the selector is in the Park or Neutral position. Therefore, the ECM "sees" 12.0V whenever the signal is high and zero volts whenever the signal is low. Based on the status of this signal, the ECM can issue commands to the IAC, for instance, to maintain an acceptable idle speed with or without the load of the transmission applied. The ECM can prevent the actuation of the TCC if the P/N signal indicates that the transmission is not in gear. Ignoring or overriding cruise control signals can also be affected by the status of this switch signal.

The Power Steering Pressure Switch status might work in a similar manner. The circuit illustrated below would serve this purpose:

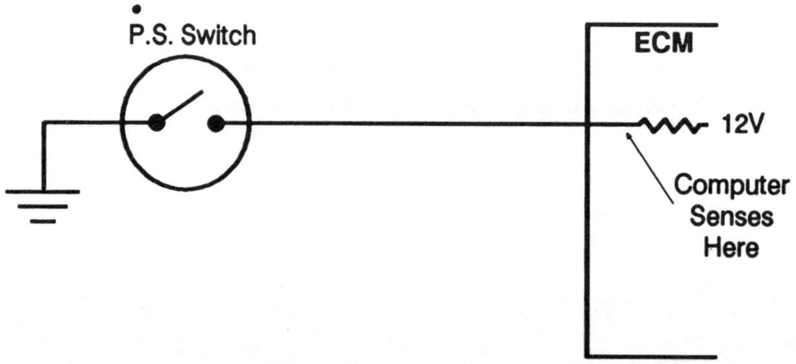

Switch Signal Input Circuits (Cont.)

Again, the ECM supplies a current limited 12.0V on the signal line. When the wheel is turned to a point which causes the power steering pressure to be higher, the switch will close, dropping the 12.0V signal to zero. The ECM detects this signal status change and issues IAC counts to compensate for the additional applied load. In some cases, the ECM might disengage the AC clutch to prevent engine stalling due to the detection of this additional load.

In the EGR Diagnostic Switch circuit illustrated below, the ECM again supplies a current limited 12.0V on the signal line. When the ECM sends a command to the EGR control solenoid, it expects the EGR solenoid to supply a vacuum to the EGR diaphragm which should cause the EGR Diagnostic Switch to close. This would pull the 12.0V signal to ground or zero volts. So, in this case the ECM doesn't use the switch signal to decide on a command, but instead, the ECM uses the switch signal to determine if the command signal sent resulted in the expected action.

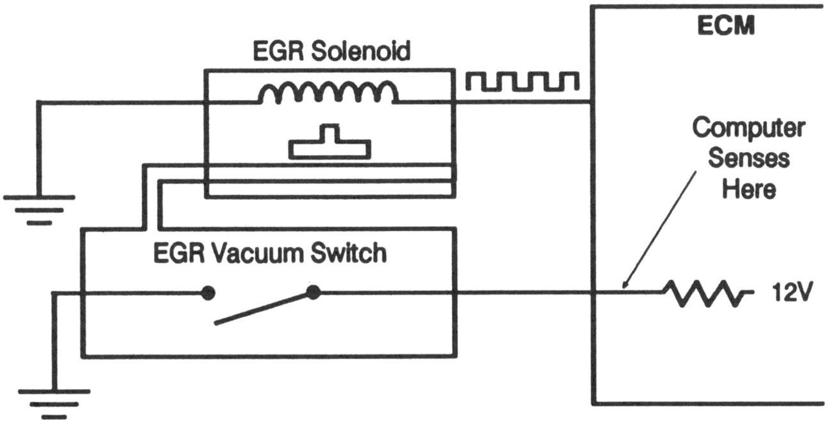

In some cases, the switch signal is the direct result of a circuit actuation. In the AC control circuit illustrated above, the ECM does not supply the voltage on the switch signal circuit for signal detection. The ECM merely "sees" the 12 volts when the AC relay is turned "ON".

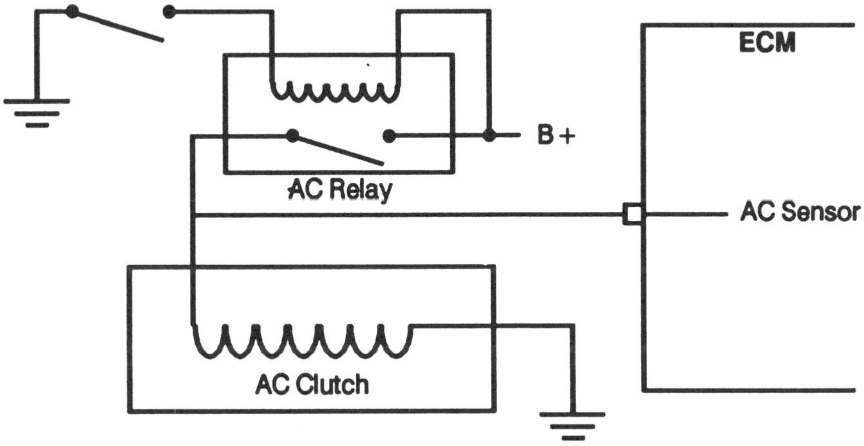

Switch Signal Input Circuits (Cont.)

The various switch signal inputs are going to be used in different roles or configurations depending on what information the ECM programmer needed for monitoring or what control status was needed for the operation(s) which were to be controlled. The signal status can be verified with a voltmeter or a logic probe at the ECM terminal by merely verifying whether the switch signal goes high or low whenever the status is changed. The signal line voltage will in most cases be 12.0V or zero volts, but, in some instances, it can be 5.0V, 8.0V or 9.0V. The voltage monitored is a function of whatever regulated voltage is applied to the signal circuit.

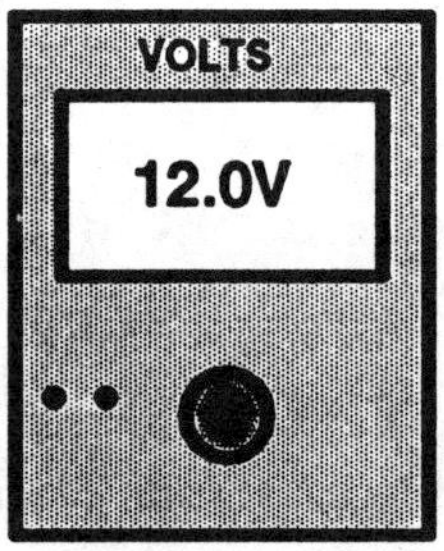

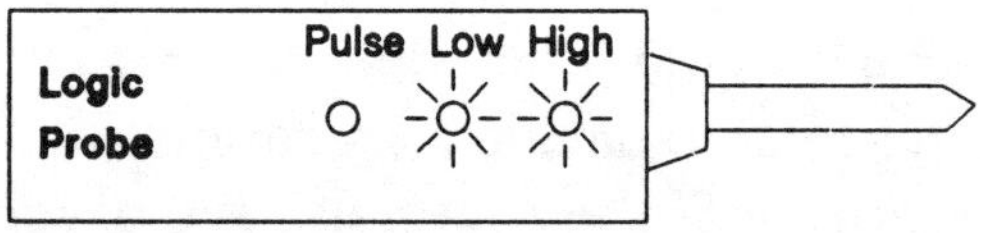

Fuel Injectors

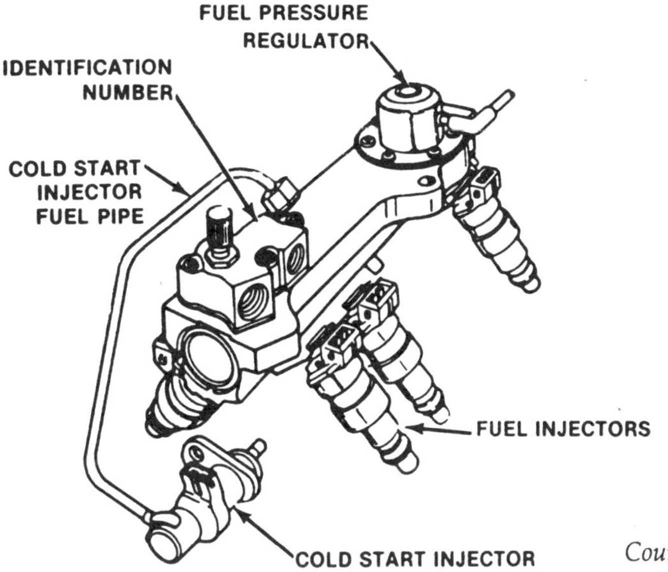

Courtesy of AC-Delco

What Are They?

The fuel injector(s) used on GM EFI engines are electric solenoid fuel injectors which can come in several configurations and quantities. The throttle body equipped engines will have one or two injectors located above the throttle bore. The PFI engines will have one injector located at each intake valve port for the number of cylinders the engine has. The fuel passage is normally closed on the injector and is designed to open, delivering fuel when the electric solenoid coil is energized.

What Are Their Jobs?

The injectors are the controlled output device used by the ECM to admit or disallow fuel entry into the engine. On a few GM engines, there may be an additional electric injector called the cold start injector. This injector is electronically controlled through a thermal time switch. The ECM does not operate the cold start injector if so equipped.

Circuit Fault Codes:

No injector codes on most GM EFI engines. The Quad 4 engine controller can set a Code 65 if the ECM senses a problem with the injector peak-and-hold circuit. All other ECMs do not specifically measure the injector current or voltage.

Fuel Injectors

Cont.

What To Watch For?

The most common injector problem has been sticky or inoperative injectors caused by restriction from foreign substances. The olefins in the gasolines are waxlike and will stick to the injector nozzles. The injector coil itself can become open or shorted which will render the injector inoperative. This has not been a common problem. The o-rings used to seal the injector fuel inlet or the intake manifold air seal can get hard or cracked causing performance problems. The injectors all have a filter screen which can also become clogged if foreign material gets into the injector.

Fuel Injector Circuits

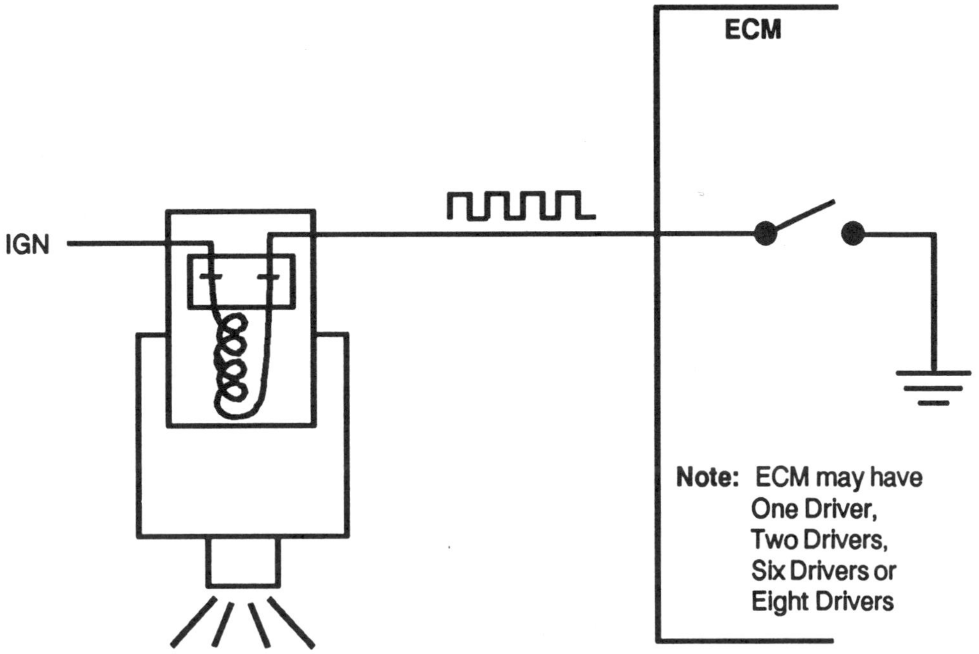

To test the injector, the technician needs only an injector pulser. Never apply 12.0V directly to the injector, this will cause the solenoid coil to burn open. Hook up the pulser, activate the injector and listen or feel for the clicking which should occur. Listening with a stethoscope to the injector being operated by the ECM will not only verify that the injector signal is being sent by the ECM, but it will also indicate whether the injector is actually operating. A clean, crisp click will be heard when listening to a good, clean injector. On throttle body systems, the actual flow of fuel through the injector can be visually verified. On PFI engines, just because the injector clicks doesn't mean that fuel has been delivered. Attaching a pressure gauge to the fuel rail, then activating the injector should produce pressure change indications if the injector activation actually results in fuel delivery.

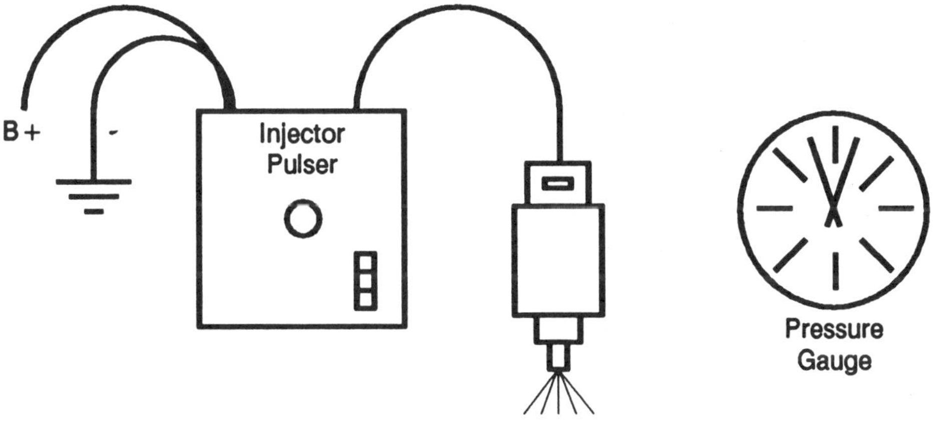

Fuel Injector Circuits (Cont.)

Testing the fuel injector signal from the ECM can be verified in several manners. Attach a Noid light, logic probe or O-scope to the injector harness. A blinking Noid light or logic probe will verify that the ECM is in fact turning the circuit ON and OFF. The O-scope will display the actual injector signal waveform. The injector pattern should look something like this:

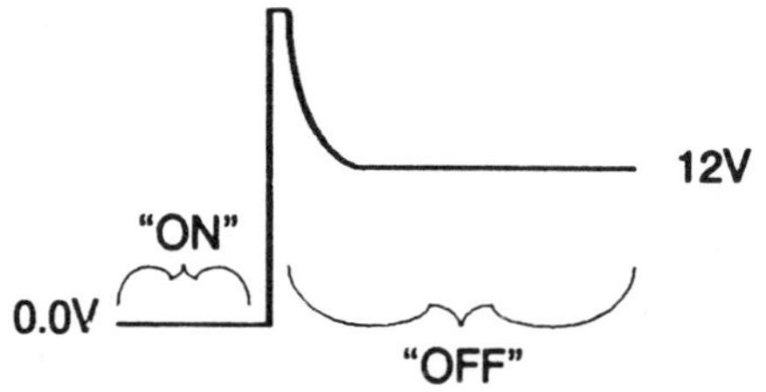

Standard Driver

Or this:

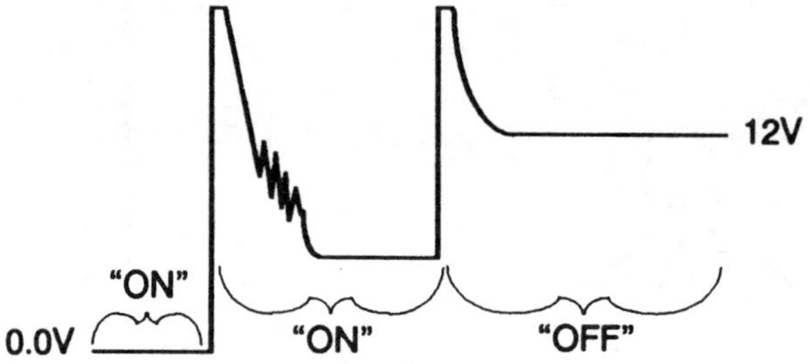

Peak and Hold Driver

The first pattern represents what you will normally see on a conventional fuel injector circuit. The second pattern represents the waveform observed on the newer low resistance injectors. These new injectors are fired with what they call a peak-and-hold driver circuit. Hence, the voltage collapse occurs twice in a very short span of time.

The scan tool will display a data stream word called injector pulse width on some GM ECMs. This pulse width is supposed to be the command sent in time to the injector(s). This pulse width command can appear as a word in the data stream even though the injector driver, the circuit or the injector is open or shorted. Merely seeing the pulse width command in the data stream does not really verify that the injector is opening and actually delivering fuel to the engine.

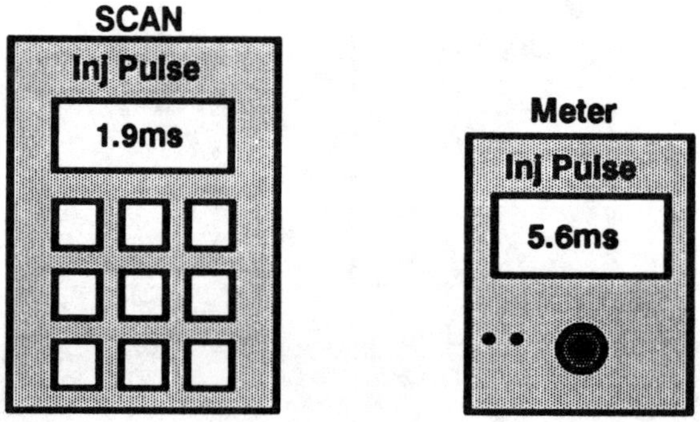

Fuel Injector Circuits (Cont.)

In summary, you can test the injector by applying a pulsed voltage directly to the injector. Just be sure the voltage applied is controlled to protect the injector coil. You can listen to the injector clicks, observe on some engines the fuel delivery or observe pressure gauge changes on others. The injector signal itself can be verified with a Noid Light or Logic Probe. The injector signal can be best observed on an O-scope. The pulse width can be measured on the O-scope pattern or you can purchase a newer style multi-meter which will measure the pulse duration for you. The number of injector drive circuits will vary from one to eight. The ECM always switches the ground side of the injector circuit. Some PFI injectors are fired in groups and some are fired individually as sequential injection systems.

Idle Air Control

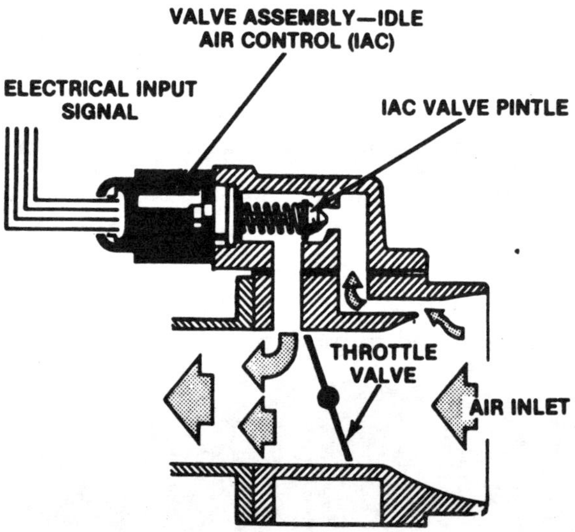

Courtesy of AC-Delco

What Is It?

The Idle Air Control or IAC is the control device used by the ECM to add or subtract the amount of air bypassed around the throttle plates to control the engine idle as various loads are applied and removed. The IAC is a four-terminal, two-coil stepper motor which is located on the throttle body assembly. Some are threaded into the throttle body. Some are bolted on to the throttle body.

What's Its Job?

The IAC motor extends or retracts a pintle in and out of the idle air bypass passage. The ECM issues commands to the IAC to position the pintle in the proper position based on the input signals received by the ECM from the various sensors and switches. Using this device allows the ECM to control the engine idle speed during all operating modes and loads.

Circuit Fault Codes:

Code 35

What To Watch For?

The IAC motor is a brushless stepper motor which theoretically means it can't burn out. The most common problem with the motor itself is sticking armature due to foreign residue buildup. The other problems that commonly occur are carbon buildup in the air passages themselves, open or shorted windings, circuit wires or ECM drivers.

Idle Air Control Circuits

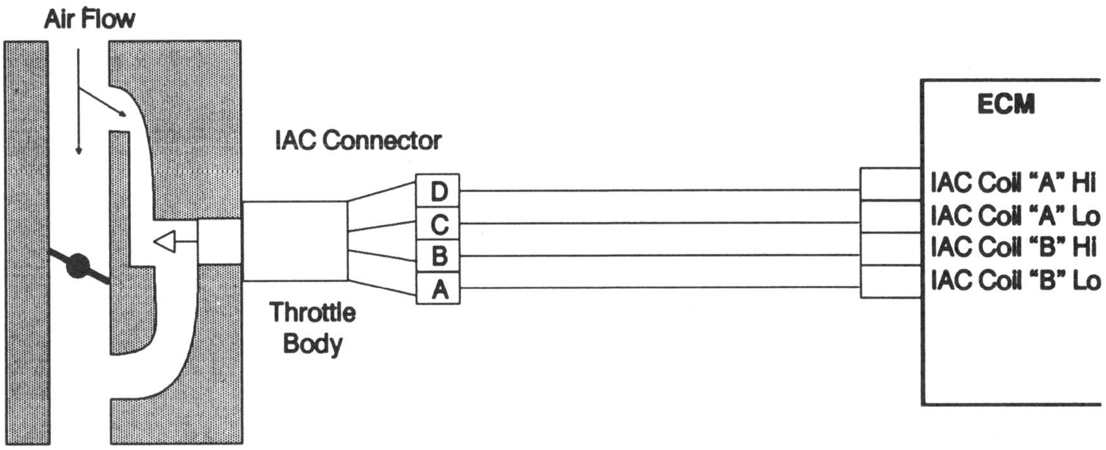

The ECM controls the movement of the IAC pintle by applying 12.0V to the IAC coils in phase steps. This application of voltage can be done in sequences which will cause the pintle to extend or retract, depending on which sequence of steps is applied. The ECM can position the pintle in any one of 256 positions. The fully extended position is called Position 0 or Step Count 0 by the ECM. The maximum retracted position theoretically would be position 255 or Step Count 255, although the pintle movement will be stopped long before the full retract counts are reached.

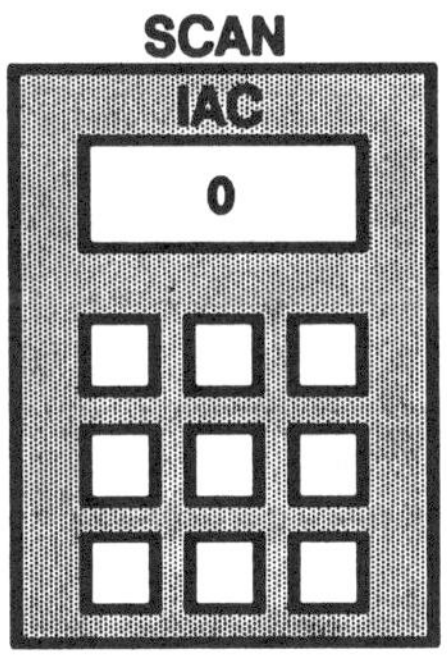

Fully Seated, No Air Bypass

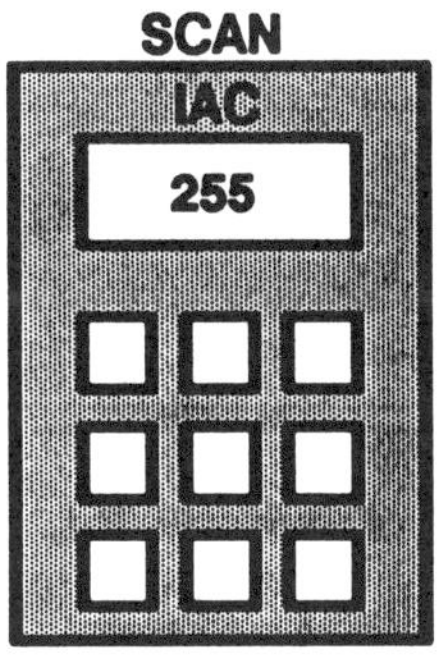

Fully Retracted, Maximum Air Bypass

If the IAC and ECM are working properly and the passages are clean, a probable normal idle step count will usually be somewhere in the 15 to 20 count, warm engine, idling in park, all accessories OFF. If the transmission gear selector is moved to DRIVE, then the ECM should see the Park/Neutral Switch change, and in anticipation of the applied load, the ECM should send additional step counts to the IAC to further retract the pintle, thereby maintaining an acceptable engine idle. This sequence of events should be repeated as each additional load is applied to the engine such as the air conditioner, power steering, etc. This sequence of events should be reversed as each load is removed.

Idle Air Control Circuits (Cont.)

The ECM's action as observed on the scan tool should be increased IAC step counts as loads are applied, and decreased step counts as loads are removed. The actual step count numbers observed will vary from one engine to another. The acceptable range for most GM engines is between 10-50 counts.

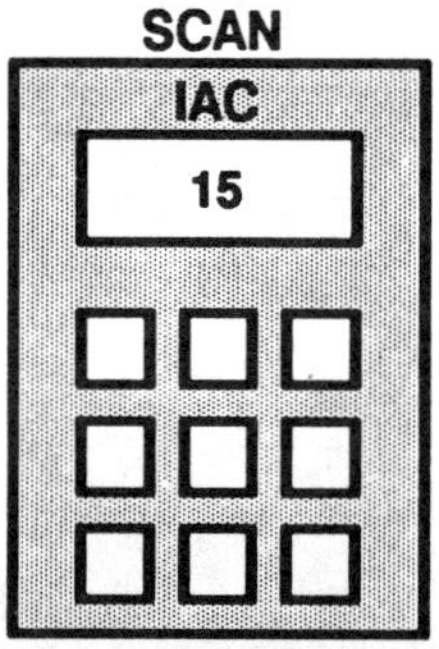

No Load Idle Count

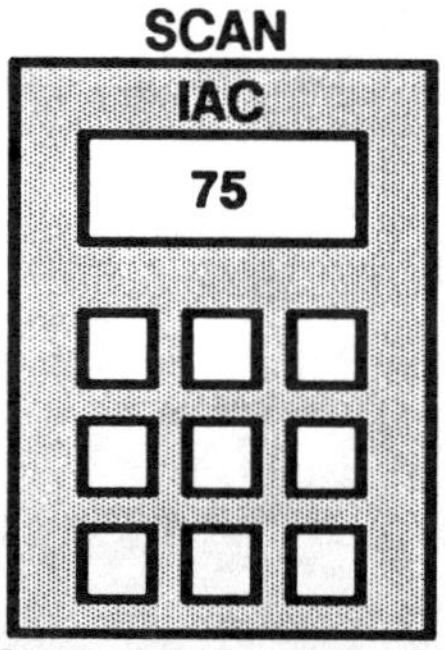

In Gear, AC On, Idle Count

The actual determination as to whether the system is operating properly is to observe the engine idle speed and quality as the engine loads are applied and removed. If the engine is idling abnormally high or low, or if the engine speed decreases too much or stalls when loads are applied, then something is not operating properly. Even though the ECM step count may be observed to be properly responding on the scan tool, if the engine idle quality is not maintained, the reading on the scan tool merely verifies that the ECM is detecting the load changes and is trying to correct the idle speed by sending the changed commands to the IAC. This does not always guarantee that the IAV pintle actually changed position properly to compensate for the changes.

If the engine is idling too high, block off the air bypass and see if the engine idle speed comes down to minimum idle. On the throttle body equipped engines, this can be done with a bypass air passage blocking tool. These are available from several different tool companies. On the PFI equipped engines, the idle air passages are not usually externally accessible. On these engines, use a stepper motor tester to send fully extended pulses to the motor. This should close off the bypass passage. If blocking the air bypass passage does reduce the abnormally high idle, then you know that the high idle was not caused by a vacuum leak or by stuck throttle plates. You also know that the IAC motor can be commanded to extend and it can close off the air passage. Now determine why the motor was incorrectly allowing too much air to be bypassed in the first place. Either the ECM was sending the wrong command, or the ECM was recieving erroneous information from an input device causing the ECM to send incorrect retract commands to the IAC. It is also possible that the ECM was attempting to extend the IAC pintle and close the air passage, but perhaps one of the ECM IAC drivers is bad or the IAC circuits could be open or shorted.

Idle Air Control Circuits (Cont.)

To test the ECM drivers and the IAC circuits, use a logic probe, high impedance test light or an IAC light style tester to monitor the voltage pulses at the IAC connector. Trick the ECM into sending voltage pulses on the four IAC circuits by jumpering the "A-B" terminals of the ALDL with the KEY ON/ENGINE OFF. An alternate method would be to force the engine to idle abnormally low while running. The ECM should be sending a steady stream of pulse commands to the IAC to correct the problem. Test each of the four IAC wires. Each should be pulsing from zero to 12.0V. If any of the four wires are not pulsing, test the four IAC driver ECM pins. If the pulses are present at the ECM, but not at the harness connector, the wire(s) are open or shorted. If the pulses are not present at any of the ECM terminals, then the drivers are bad. It is unlikely that all four IAC ECM drivers would be bad at the same time. If no pulses are present at all of the ECM terminals, then the ECM is failing to issue pulses to the IAC circuits entirely.

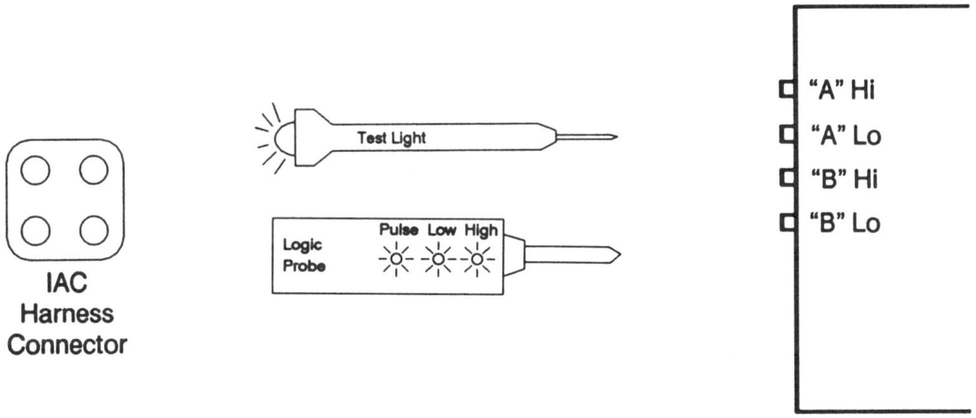

In summary, the ECM's job is to issue phased step pulses to the IAC to extend or retract the pintle. The IAC's job is to physically open or close the idle air passage according to the issued step count command. The idle air passage's job is to bypass air from above the throttle plate to below the throttle plate so acceptable engine idle quality can be maintained with various engine loads applied and removed. Watch out for plugged air passages. Watch out for sticky IAC motors. Watch out for binding throttle plates. Be prepared to drive the motor in and out with a tester that you can control and to monitor the IAC step pulses manually rather than merely watching the step count changes on the scan tool.

Electronic Spark Timing and Electronic Spark Control

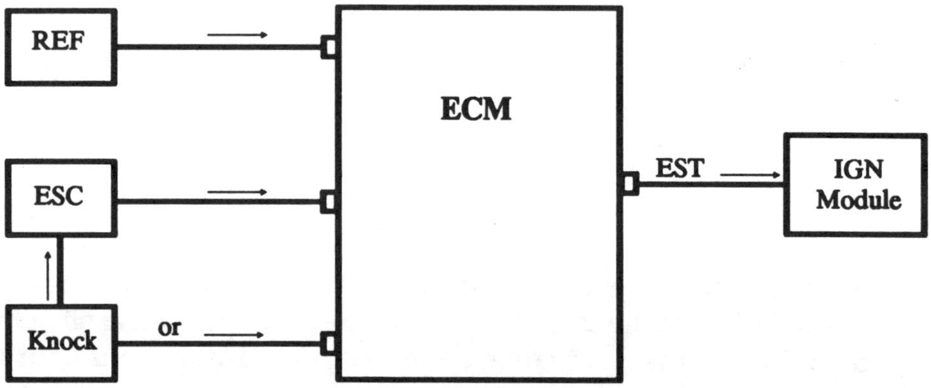

What Are They?

Electronic Spark Timing or EST and Electronic Spark Control, or ESC, are not devices controlled by the ECM. They are outgoing signals sent by the ECM to the ignition module in order for the ECM to be able to control the timing advance and spark retard of the engine. All GM EFI engines have EST. All turbo-charged GM EFI engines have ESC as do some non-turbo applications.

What Are Their Jobs?

The EST signal is the time-shifted reference signal sent to the module to fire the ignition in advance of base timing. The purpose of EST is to allow the ECM to control the application of timing advance electronically since this represents a faster and more accurate method of timing control. The ESC modification of the timing control signal occurs internally within the ECM when the related ESC hardware informs the ECM that detrimental detonation or spark knock is occurring. The net effect is that the ESC can represent a retard modification of the time advanced EST signal.

Circuit Fault Codes:

Code 42 or Code 43

What To Watch For?

Open or shorted EST or Bypass lines, inoperative ECM timing management sub-section or inoperative ESC hardware or erroneous knock signals from the knock sensor due to engine vibrations other than actual spark knock.

EST and ESC Circuits

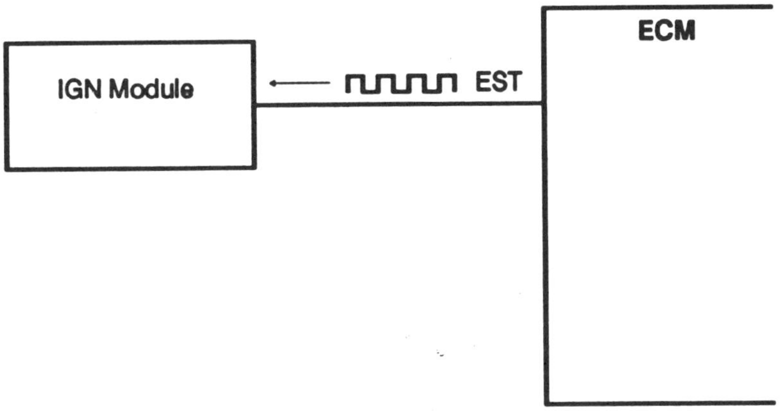

Testing the EST and ESC circuits is discussed in detail in Section II of this manual. The essential factor involved is the time shifting of the EST control signal from the ECM to the ignition module. Observing that signal on an O-scope will reveal that it is a simple DC square wave. The time shifting effect is difficult to observe on this waveform because of the referencing difficulty. The end result of the time shift factor can be observed using an advance control timing tester to verify that the ECM can actually change the timing.

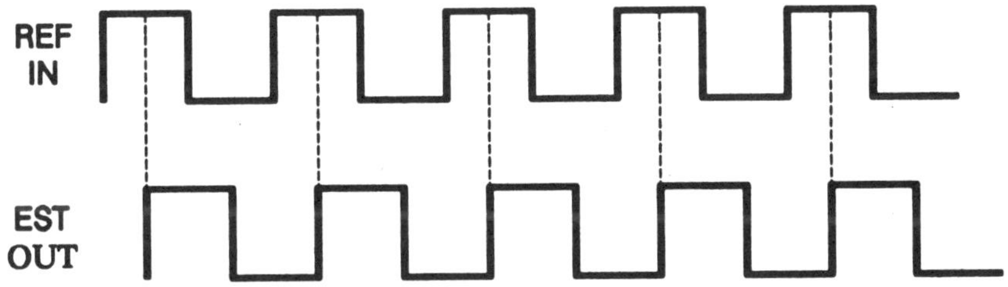

Exhaust Gas Recirculation Valve

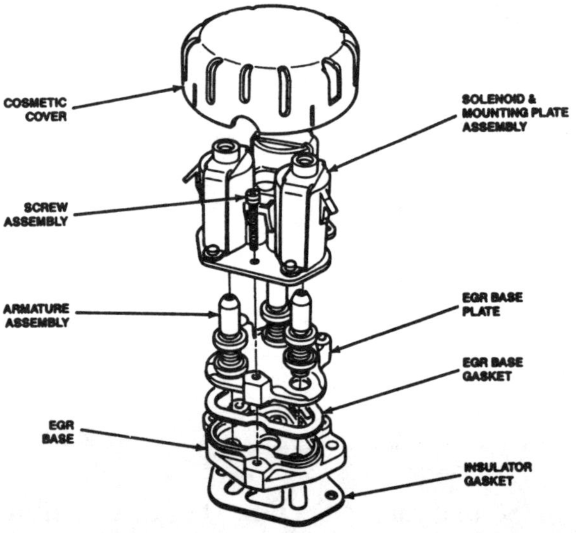

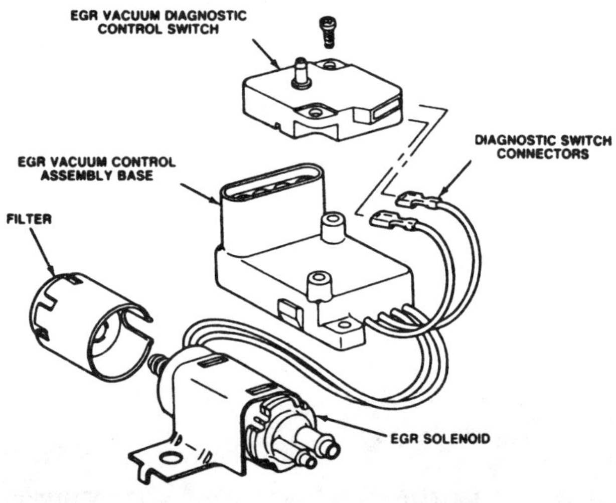

Courtesy of AC-Delco

What Is It?

The EGR valves used on GM EFI engines can be of several different varieties. The basic EGR valve is a simple backpressure type of vacuum operated EGR valve. The ECM uses a simple solenoid to control this type of valve.

Some engines will use a standard EGR valve with a pulse width modulated solenoid supplying varying amounts of vacuum based on a control signal from the ECM. Some of these solenoids have a thermal sensing switch or a vacuum sensing switch contained within the solenoid housing.

Some GM engines use an electronic solenoid assembly where the pulse width modified control solenoid is built into the EGR valve assembly. This unit also has a potentiometer type EGR position sensor to inform the ECM of EGR position.

The newest GM EGR valve is called the Digital EGR and it is fully electronic. This valve contains three solenoid coils which operate three EGR rods for three EGR passages. The ECM can send commands to this assembly in varying combinations to achieve the proper amount of exhaust gas recirculation for various engine conditions. There is no vacuum signal needed for this valve's operation.

What's Its Job?

The EGR valve and or the related solenoid(s) are the control devices used by the ECM to prevent the formation of NOx by admitting exhaust gases into the combustion chamber to lower the combustion temperature. This also serves the dual purpose of preventing engine detonation.

Exhaust Gas Recirculation Valve

Cont.

Circuit Fault Codes:

Code 31, Code 32, Code 53, Code 63, Code 64 or Code 65

What To Watch For?

Since most of the EGR systems used on GM EFI engines are electro-mechanical systems, the most common problems will be with the mechanical part of the system such as vacuum tubing, etc. The EGR passages are also commonly known for carbon buildup which will render the system ineffective. The solenoids themselves are subject to blocked passages, open or shorted coils, or dirty switch contacts on the units with diagnostic switches contained within.

EGR Circuits

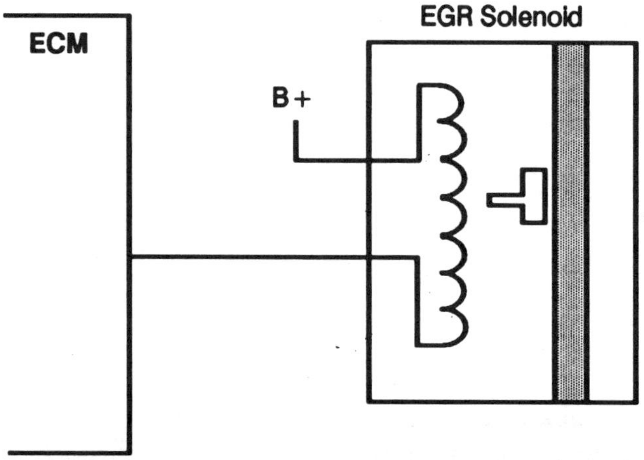

To test the EGR system involves first identifying what type of system is being used on the vehicle. The net result of a properly operating EGR valve is no fault codes, reduced NOx levels in the exhaust and no engine detonation under acceleration.

The amount of NOx present in the exhaust stream is very minute and the gas is very volatile. You cannot measure NOx at the tailpipe with a standard IR. The exhaust gas has to be captured and measured in a labratory under very tightly controlled labratory conditions.

A road test will reveal whether or not the engine knocks under cruise condition, however, spark knock can be caused from conditions other than an inoperative EGR system so that's not conclusive either.

The presence of a fault code does not necessarily mean that the EGR is inoperative. The absence of a fault code does not necessarily mean that the EGR is operating, so how can you tell if the EGR is OK?

First, after identifying the system used, identify the normal proper operating conditions of the EGR. You never need or want the EGR to be open at idle, WOT or acceleration. The EGR should be open only under steady cruise operation. In most cases you can manually operate the EGR valve while idling the engine. If the engine stalls or drops RPM significantly when you open the valve, the EGR passages are probably clear. If you can activate the EGR with a hand vacuum pump, the EGR diaphragm is OK. The backpressure type of valves will only open if the engine speed is increased enough to create sufficient exhaust backpressure. If the EGR passages are clear, and the EGR valve can be manually activated, then the next step is to verify that the EGR system electronics are operating.

EGR Circuits (Cont.)

The scan tool can display EGR electronic status in several different ways, depending on which system is used. The simple solenoid equipped systems will show either ON or OFF when the EGR data stream word is selected. Be careful here, because on some systems the EGR solenoid is mechanically blocking the vacuum supply to the EGR valve when the solenoid is electrically ON. Determine whether your scan tool is displaying the word for system electrical or mechanical status. Some systems will display EGR status as a percent of EGR flow or blockage. This could be a number between 0-100%. The trick here is to remember that the displayed scan tool reading represents the amount or the status of the ECM's command to the EGR control solenoid. In order to be sure the EGR valve is actually opening, you have to verify that the vacuum is being supplied to the EGR which will open the valve.

SCAN

SCAN

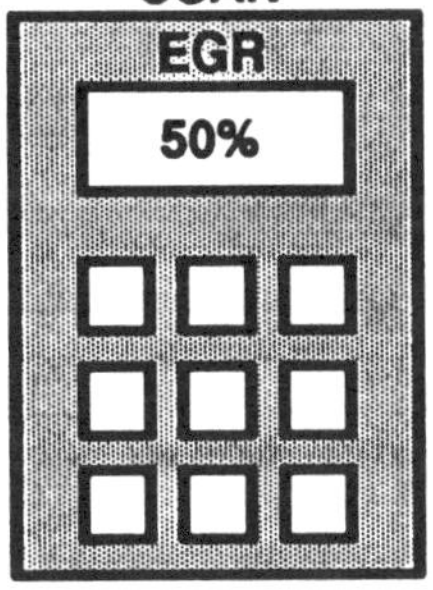

On the electronic EGR valve with the potentiometer type of position sensor you can check the position sensor signal line voltage and make a potentiometer sweep test to check for cracks or worn spots on the sensor resistance substrate.

Volts

Signal Voltage

Ohms

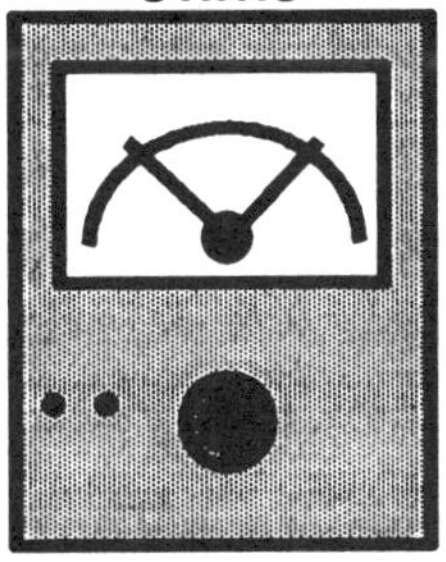

Sweep Test

The Digital EGR is strictly electronic. You cannot manually push or pull on anything to open any of the three EGR ports on the valve. You can check the continuity and resistance of each of the three solenoid windings and you can apply voltage momentarily to each solenoid coil using jumper leads. Each coil operates a rod which retracts, opening the passage when the solenoid is energized. There is a small, a medium and a large EGR passage opening under the valve. The ECM has three driver circuits which operate the three solenoids. You can monitor each of the three circuits as you increase RPM and verify ECM driver operation. This

EGR Circuits (Cont.)

particular setup features ECM code detection for the current draw by each solenoid. That is why GM has a code (63, 64, 65) assigned to each solenoid circuit. This is a departure from the norm for GM. They do not normally "sense" the current draw on a driven circuit.

In summary, the verification process for GM EGR systems varies depending on which EGR system is being tested. The scan tool readings will tell you that the ECM is recognizing the inputs telling it to open the EGR valve and send the ON command to the EGR solenoid. To be sure the valve is actually operating, you have to physically verify that the valve can open and that the passages are clear. Then electrically verify that B+ power is available to the solenoid and that the ECM driver(s) can and actually do complete the circuit(s) to ground, opening the valve or vacuum passage controlled by the solenoid.

Output Solenoids

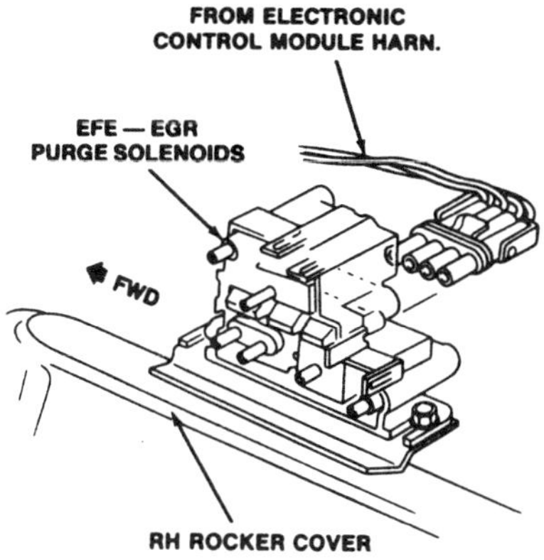

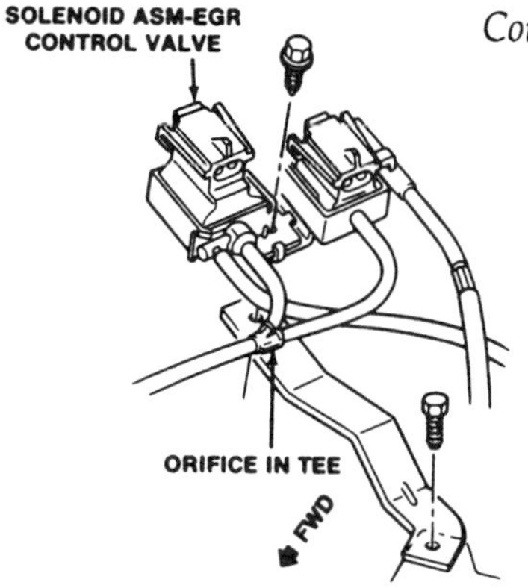

Courtesy of AC-Delco

What Are They?

The output solenoids used on GM EFI engines will vary according to application and engine requirements. The solenoids are the control devices used by the ECM to actuate various subsystems. Such systems as the Torque Converter Clutch, the air management control valves, the turbo wastegate or the canister purge valve are commonly used output solenoid controlled systems. Some GM vehicles use ECM controlled solenoids to operate the cruise control servo.

What Are Their Jobs?

The jobs that each perform relate to the named purpose of each solenoid. The TCC solenoid is designed to reduce torque converter slippage, thereby reducing engine RPM and fuel usage during cruise conditions. The air management valves and the canister purge solenoid are both designed to reduce emissions. The turbo wastegate solenoid controls the turbo wastegate to prevent the engine from overboost which could result in engine damage. The cruise control servo solenoids apply and vent vacuum to the cruise control transducer.

Circuit Fault Codes:

Code 31, Code 61

Output Solenoids

Cont.

What To Watch For?

In almost all cases, the ECM does not supply B+ to any of the solenoids. The ECM usually only completes the circuit path to ground by turning ON a driver transistor which closes the ground path. The driven solenoids can easily burn out the ECM driver if the solenoid is shorted. The trick to understanding GM's self-diagnostic philosophy concerning solenoids is to realize that GM usually doesn't fault code the solenoid driver circuit itself. They usually fault code some sensor which the ECM is using to monitor and verify that the solenoid has been activated and effective at the job it is designed to do. Some of the newer GM systems are now beginning to monitor the driven circuit itself, such as the Digital EGR valve. There are also some new codes showing up related to the ECM Quad drivers which are used to control these various output solenoid devices.

Output Solenoid Circuits

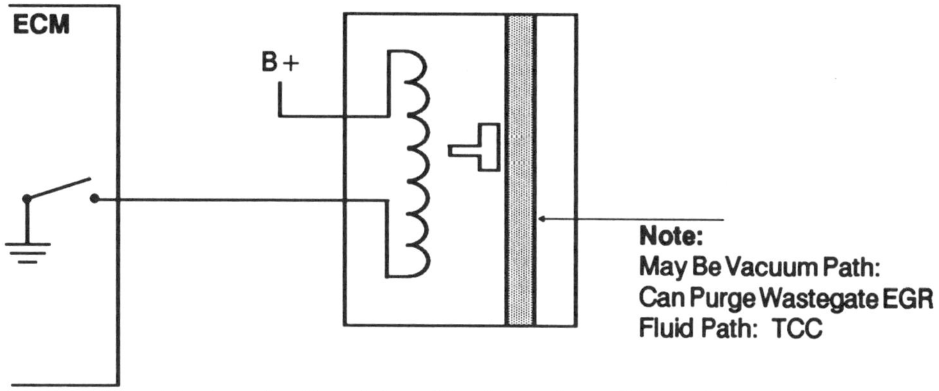

To test the output solenoid circuit, first identify the solenoid used and determine what action or actions it should implement when electrically activated. For instance, the TCC solenoid should significantly reduce cruise engine RPM when activated, thereby reducing engine fuel consumption. To test the TCC solenoid, test drive the vehicle under cruise conditions. If the TCC is being electrically turned ON, then the cruise RPM at 55 MPH should be around the 2000-2500 RPM. If you observe the TCC status light or display on the scan tool, the TCC should be ON when you are cruising and it should turn OFF if you tap the brake pedal or slightly engage the emergency brake. It should also turn OFF if you accelarate hard. The RPM ranges you will observe will vary for different vehicles, but you will see the normal ranges after you have driven several equipped with TCC. Also, be on the lookout for additional switches which might be in the circuit such as gear switches, etc. The switches may be in the circuit to prevent solenoid actuation under the wrong conditions or the switches may be directly inputted to the ECM to ensure proper condition satisfaction before the ECM will issue an ON command.

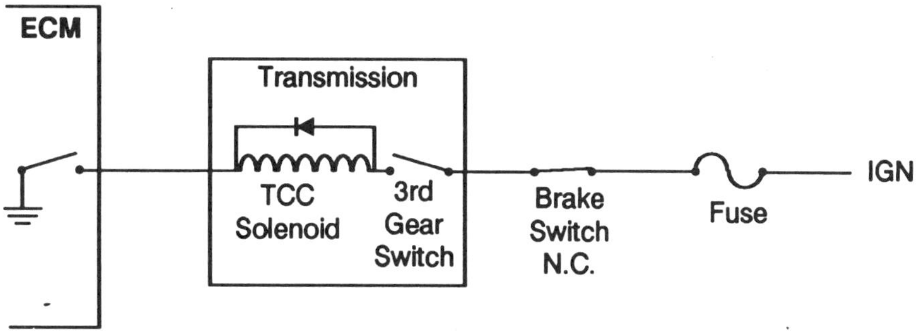

To test a different kind of solenoid such as the Air Management valve, you have to manually activate the valve or valves using a jumper wire to activate the valve gate. Then physically verify if the air path has changed to the correct channel. The valve itself can have a stuck door or perhaps the valve gate is leaking air past the gate when the solenoid is electrically OFF. Most GM air management valves are direct solenoid operated gate valve rather than the solenoid operating a vacuum channel to the remote air gate valve.

Output Solenoid Circuits (Cont.)

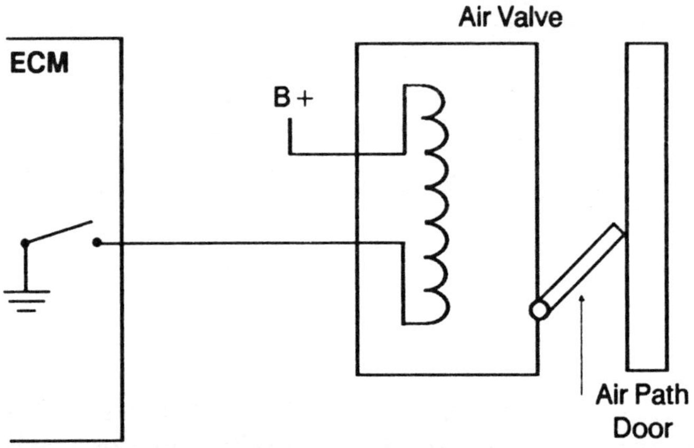

This procedural philosophy can be applied to each and every type of GM output solenoid control system. The role that the solenoid plays will vary according to the job or jobs it was designed to perform. The mechanical verification is essential to complete determination of the system operation.

Electrically verifying the circuit itself is a simple matter of using a voltmeter, test light or logic probe to test the B+ supply to the solenoid and the voltage on the driven side of the solenoid. You can also measure the solenoid coil resistance to verify continuity and coil condition. The diode is used to protect the EMC when the coil field collapses. The diodes are hidden, but they are almost always present. An open/shorted diode can cause EMC troubles or intermittent code(s).

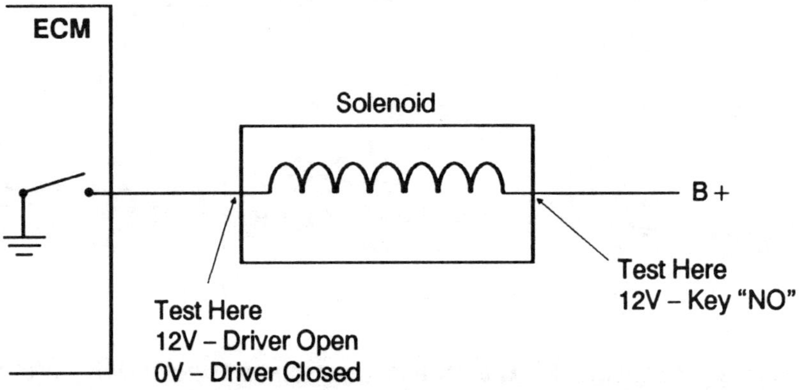

In summary, observing the status of the solenoid with the scan tool will verify whether or not the ECM is commanding the circuit to be ON or OFF. The actual electrical verification must be done manually with some type of voltage indicator. The physical action implemented by the solenoid must be cross checked using whatever method is appropriate to verify that the action has taken place. Remember, most of the solenoid drivers will be turned electrically ON when you jumper the "A-B" terminals of the ALDL with the ignition key ON and the engine OFF.

Output Relays

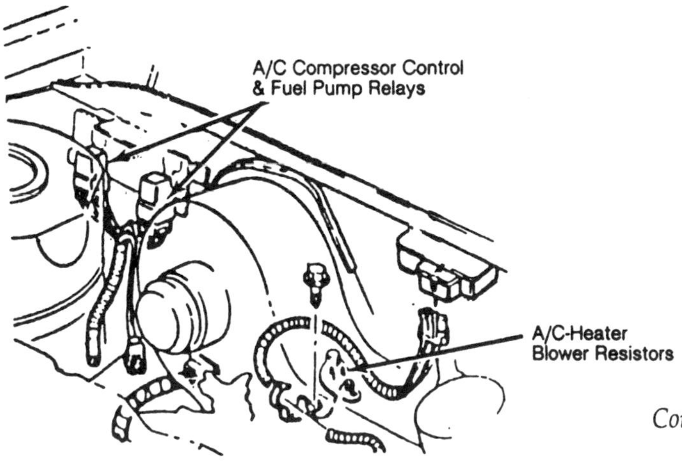

Courtesy of AC-Delco

What Are They?

The output relays controlled by the ECM are found in various configurations. The relay operating philosophy will vary from one relay to another. Relays may be used to directly control such things as the fuel pump, the AC clutch or the cooling fan.

What Are Their Jobs?

The job of each relay is specified by its name. The job itself may be one of direct control, or it may be designed as an override control in such cases where the main control circuit switch does not operate due to parameter limitations.

Circuit Fault Codes:

Code 36

What To Watch For?

The relays controlled by the ECM will in most cases be supplied a ground path for the pull-in coil by the ECM. Some relays, such as the fuel pump relay may be supplied B+. Watch the circuit carefully, some relays will have more than one pull-in coil and may supply battery connection to more than one device or circuit. GM relays, like all relays are only as good as the duty cycle capacity they are manufactured to provide. Remember, they are designed by engineers and then purchased from the low bidder.

Output Relay Circuits

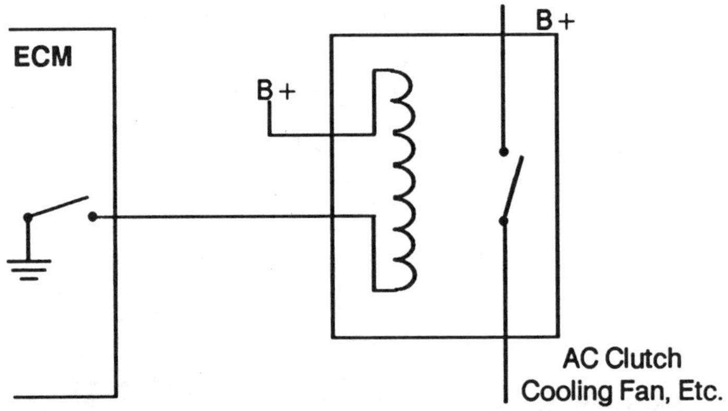

The ability to test the relay circuit depends greatly on the proper identification of the relay terminals and the assignment to each terminal. The electrical testing can then be performed with a simple voltmeter, test light or logic probe. For instance, in the circuit above, the ECM supplies the ground path for the pull-in coil. The relay should then close the contacts completing the B+ supply to the load.

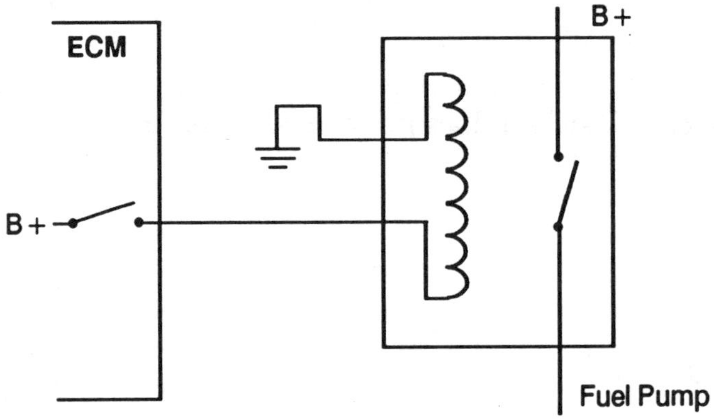

In the circuit above, the ECM does just the opposite. It supplies B+ to the relay pull-in coil so that the relay contacts can close completing the B+ supply voltage to the load. Remember that the circuit to the load or to the relay can also contain supplemental switches to ensure proper circuit operation or to interrupt circuit operation if all the necessary parameters are not met.

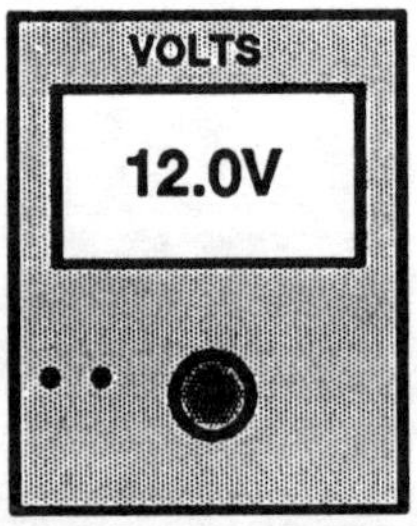

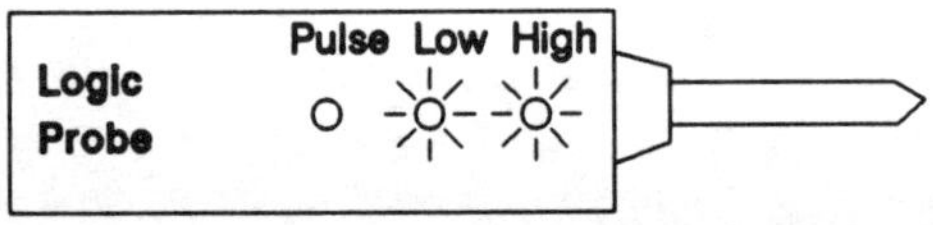

Review of Key Concepts

- **Port fuel inputs can consist of thermistors, potentiometers, voltage generators, frequency generator, piezo resistive thick film pressure transducers, and switch circuits.**

- **The technician can test all of these devices if he understands how they work and which tester or testers best analyze the circuit or device.**

- **Port fuel outputs can consist of simple solenoid coils, pulse width modulated solenoids, and alternating polarity sequential stepper motor commands. In addition, the ECM outputs a word stream call "serial data".**

- **The technician can test all of these devices if he understands how they work and which tester or tools best analyze the circuit or device.**

Table of Contents

Section IV

Section IV Outline

IV. GM ECM Common Pin Assignments

A. GM EFI engine controllers have some common pin assignments for engine controller types.
   1. There are six different types of GM EFI ECMs with common harness connector types.

B. As the model years progressed, the ECM fault code libraries continued to expand.

C. In section III, the individual circuits are illustrated. Using the pin identifiers in section IV, the technician can see the whole circuit, including where each wire enters or exits the ECM.
   1. These pin identifiers are ideal for using a test probe interface.
   2. These pin identifiers can also be applied to circuit testing with the scanner.

GM ECM Connectors

The skilled electronic technician prepares himself to troubleshoot the circuits and systems of the device he intends to service by first gathering the necessary documentation on that device. You would quickly lose confidence in a serviceman's capability if he attempted haphazardly to disassemble and repair a valuable electronic component of yours without first having the necessary schematics and service information in hand.

All of the previously published Computerized Fuel Injection diagnostic charts available to the service technician are written based on the premise that the service technician doesn't want to know or have the capability of learning the essential operating characteristics of the various circuits and devices utilized in the different forms of Computerized Fuel Injection Systems.

Each diagnostic chart, or "tree" as they are sometimes called, is designed to lead the technician through a series of steps which, if carefully followed, should lead even the totally uninformed serviceman to the right component for replacement. This "tell them nothing" philosophy has led to a great deal of disappointment and frustration for both the service technician and the vehicle owner. The following situations can, and have occurred frequently:

1. The technician uses the wrong chart and reaches an erroneous conclusion. Too many charts for all models.

2. The part replaced does not solve the problem and cannot be returned. Who pays for that part?

3. The right chart was used, but the wrong diagnosis occurred because the technician did not know how to correctly interpret the test results.

Again, these situations cause an expensive loss of time and incorrect parts replacement. The following pages contain a set of GM ECM pin charts. These charts have been collated from a large number of electrical schematics in order to condense and simplify the technician's task. These charts illustrate to the trained technician that there are some "common connections" chosen by GM's design engineers which will serve the technician during circuit troubleshooting. These charts can be utilized along with the circuit and component descriptions to complete the circuit and utilize direct circuit testing methods which are commonly used by electronic technicians in other areas of the industry.

GM ECM Connectors

Fuel Injected Engines

There are six different GM harness connectors utilized by GM for its fuel injected engine families. They are illustrated below and on the following pages.

Type One:

2 Connectors – 1 Black and 1 White

White connector contains 24 electrical terminals.

Black connector contains 22 electrical terminals.

Note: These connectors are called "edgeboard" connectors. The wire terminals used are called "pack-con" and can be released individually from the connectors with the proper tool.

Black Connector

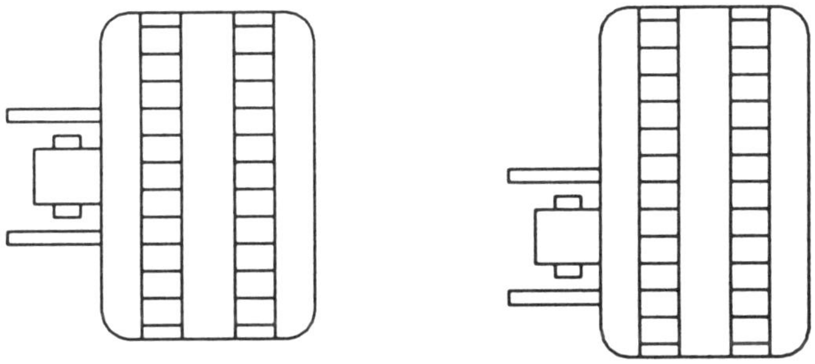

White Connector

Type Two:

2 Connectors – 1 Black and 1 White

White and Black Connector each contain 24 electrical terminals.

Note: These connectors are called "Header" style connectors. They use male and female style pins. They can be removed individually with the proper tool.

White 24 Pin Connector (W)

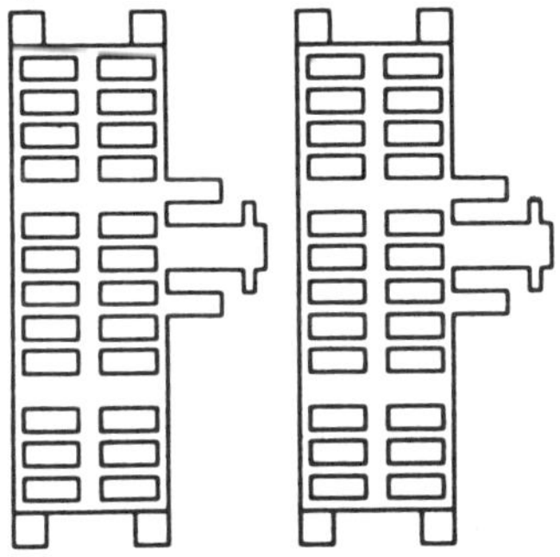

Black 24 Pin Connector (W)

GM ECM Connectors (Cont.)

Type Three:

2 Connectors – No set color standard

One connector contains 24 pins and is lettered A–B.

One connector contains 32 pins and is lettered C–D.

Note: These connectors are also called "Header" style connectors.

24 Pin A–B Connector

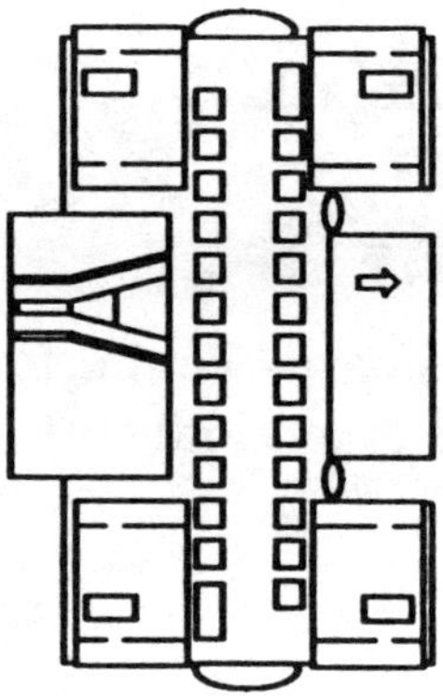

36 Pin C–D Connector

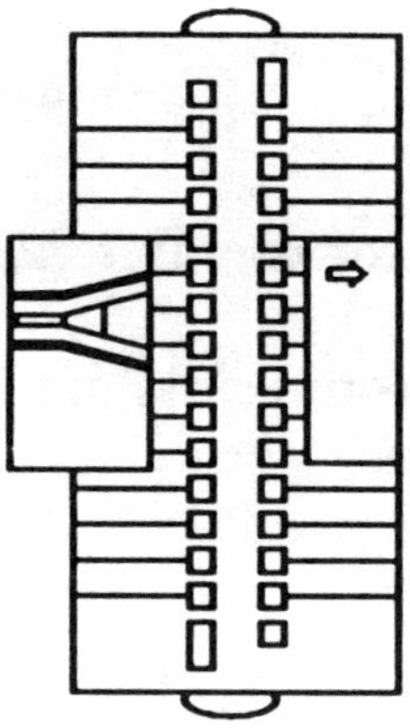

Type Four, Five, and Six:

3 Connectors – 2 Black and 1 Yellow or Orange.

One Black Connector contains 24 Pins and is lettered A–B.

One Black Connector contains 32 Pins and is lettered C–D.

One YEL/ORN Connector contains 32 Pins and is lettered E–F.

24 Pin A–B Black Connector

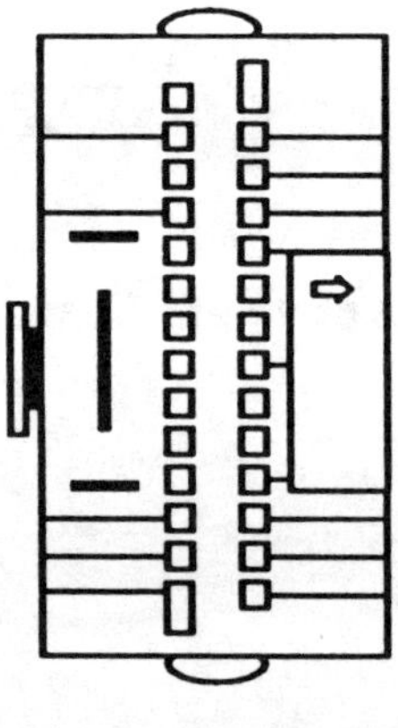

32 Pin C–D Black Connector

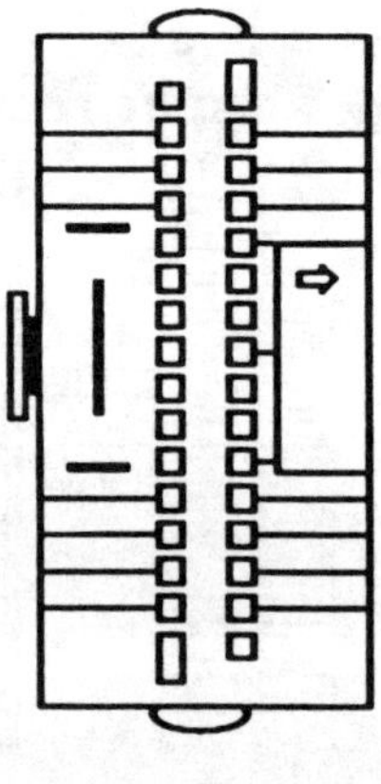

32 Pin E–F YEL/ORN Connector

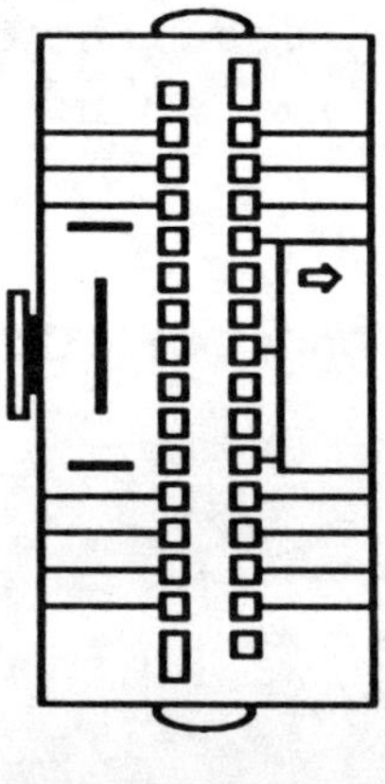

Pin Identifier

Type One ECM Pin Connectors 22 and 24 Pins

Black 22 Pin Connector

Wire Color	Circuit I.D.	Pin	Pin	Circuit I.D.	Wire Color
GRY	EGR	22	1	Crank	PPL/WHT
TAN	Ambient Temp	21	2	HEI Ref.	PPL/WHT
LT GRN	MAP	20	3	HEI Ref. Grd.	BLK/RED
WHT	EST	19	4	Coolant Sensor	YEL
LT BLU/BLK	IAC Coil "A"	18	5	TPS Sig.	DK BLU
LT BLU/BLK	IAC Coil "A"	17	6	Pwr. Strg.	LT BLU/ORN
BLK/PNK	Air Divert	16	7	AC Clutch Relay	DK BLU
TAN	Oxy Sensor Grd.	15	8	Oxy. Sensor	PPL
BRN	Air Switch Sol.	14	9	Cooling Fan or or Can. Purge	LT BLU/BLK DK GRN/YEL
LT GRN/BLK	IAC Coil "B"	13	10	EST Bypass	TAN/BLK
LT GRN/WHT	AC Coil "B"	12	11	Coolant & TPS Ground	BLK

Pin Identifier

Type One ECM Pin Connectors 22 and 24 Pins

White 24 Pin Connector

Wire Color	Circuit I.D.	Pin	Pin	Circuit I.D.	Wire Color
	Spare	24	1		Spare
DK BLU/WHT	E-Cell	23	2	VSS	BRN
DK GRN	4th Gear Switch	22	3	Diagnostic	WHT/BLK
DK GRN/WHT	AC Clutch	21	4	ESC	BLK
BRN/WHT	CEL	20	5	P/N	ORN/BLK
TAN/BLK	TCC	19	6	Dual Inj. Sol.	LT BLU
DK GRN/WHT	FP Relay	18	7	Serial Data	ORN
TAN/WHT LT BLU	Bat. V Mon.	17	8	Inj. #1	BLU
PNK/BLK	Sw. Ign.	16	9	Inj. #2	LT GRN
ORN	Battery	15	10	Battery	ORN
BLK	MAP Grd.	14	11	5 Volt Ref.	GRY
BLK\WHT	ECM Grd.	13	12	ECM Ground	BLK/WHT

Pin Identifier

Type Two ECM Pin Connectors 24 and 24 Pins

White 24 Pin Connector (W)

Wire Color	Circuit I.D.	Pin	Pin	Circuit I.D.	Wire Color
TAN	Injector	W1	W24	Fuel Pump Rly	DK GRN
DK GRN	AC Clutch	W2	W23	VSS Opt.	BRN
RED	VSS 2000 PPM	W3	W22	VSS 4000 PPM	RED/WHT
	Cruise	W4	W21	Cruise	
BLK\WHT	PS Sig.	W5	W20	Cruise	
WHT	EST	W6	W19	EST Bypass	TAN/BLK
LT BLU/WHT	IAC "B" Hi	W7	W18	P/N	ORN/BLK
LT GRN/BLK	IAC "B" Lo	W8	W17	IAC "A" Lo	LT BLU/BLK
LT GRN/WHT	IAC "B" Hi	W9	W16	Sw. Ign.	PNK/BLK
ORN/BLK	Battery	W10	W15	Battery	ORN/BLK
GRY	5 Volt Ref.	W11	W14	MAP MAT Grd	BLK/ORN
BLK/WHT ECM	Ground	W12	W13	ECM Ground	BLK/WHT

Pin Identifier

Type Two ECM Pin Connectors 24 and 24 Pins

Black 24 Pin Connector (B)

Wire Color	Circuit I.D.	Pin	Pin	Circuit I.D.	Wire Color
ORN	Serial Data	B1	B24	AC Signal for Fan	LT BLU/BLK
PPL	Oxy Sensor	B2	B23	Oxy Sensor Grd.	TAN
DRK BLU	AC Clutch Rly	B3	B22	SES Light	BRN/WHT
Spare		B4	B21	Eng. Cool Fan	DK GRN/WHT
TAN	MAT	B5	B20	MAP	LT GRN
YEL	Internal VSS	B6	B19	TPS	DK BLU
TAN/BLK	TCC or Shift	B7	B18	Cruise	
YEL	Coolant	B8	B17	Spare	
PPL/WHT	HEI Ref.	B9	B16	HEI Ref. Grd.	BLK/RED
	Cruise	B10	B15	Diagnostic	WHT/BLK
	Cruise	B11	B14	Cruise	
BLK	Coolant-TPS Grd.	B12	B13	Internal VSS	

Pin Identifier

Type Three ECM Pin Connectors 24 and 32 Pins

Throttle Body Injection

Wire Color	Circuit I.D.	Pin
DK GRN/WHT	Fuel Pump	A1
[1]DK BRN	AC Cluch	A2
[2]GRY/RED	Can. Purge	A3
GRY	EGR Conrol	A4
BRN/WHT	CEL or SES	A5
PNK/BLK	Sw. Ign.	A6
TAN/BLK	TCC	A7
ORN	Serial Data	A8
WHT/BLK	Diagn.	A9
BRN	VSS	A10
BLK	5V Return	A11
BLK/WHT	System Grd.	A12

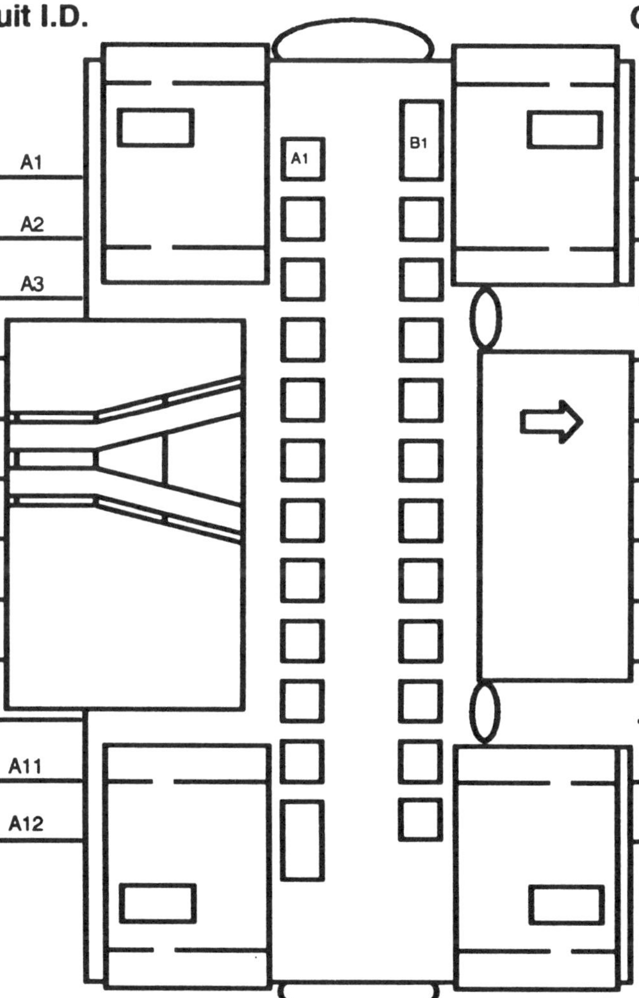

Pin	Circuit I.D.	Wire Color
B1	Battery	ORN
B2	Voltage Monitor	TAN/WHT
B3	HEI Ref. Grd.	BLK/RED
B4	Not Used	
B5	HEI Ref.	PPL/WHT
B6	Not Used	
B7	ESC	YEL/BLK
B8	AC Signal	DK GRN
B9	EGR Switch	BLU
B10	P/N	ORN/BLK
B11	Not Used	
B12	Not Used	

24 Pin A-B Connector

[1] Used for Can. Purge 2.0L
Used for TCC 2.5L Truck

[2] Used for AC Rly 2.0L
Used for EGR 2.5L Truck

Pin Identifier

Type Three ECM Pin Connectors 24 and 32 Pins

Throttle Body Injection

Wire Color	Circuit I.D.	Pin	Pin	Circuit I.D.	Wire Color
[3]GRN/WHT	Cooling Fan	C1	D1	System Grd.	TAN/WHT
[4]BRN	Air Sw. Sol 4.3	C2	D2	MAT Grd.	BLK
LT GRN/BLK	IAC "B" Lo	C3	D3	Not Used	
LT GRN/WHT	IAC "B" Hi	C4	D4	EST Control	WHT
LT BLU/WHT	IAC "A" Hi	C5	D5	5V Bypass	TAN/WHT
LT BLU/BLK	IAC "B" Lo	C6	D6	0^2 Grd.	TAN
LT BLU	4th Gear 4.3L	C7	D7	0^2 Signal	PPL
DK GRN	PS Sw.	C8	D8	Not Used	
PPL/WHT	Crank	C9	D9	Not Used	
YEL	Coolant	C10	D10	Not Used	
LT GRN	MAP	C11	D11	Not Used	
TAN	MAT	C12	D12	Not Used	
DK BLU	TPS	C13	D13	Not Used	
GRY	5V Rev	C14	D14	Inj. B 2.8L 4.3L	LT GRN
LT GRN	Inj. B 4.3L	C15	D15	Inj. A 4.3L	LT GRN
ORN	Battery	C16	[5]D16	Inj. A 4.3L	LT GRN

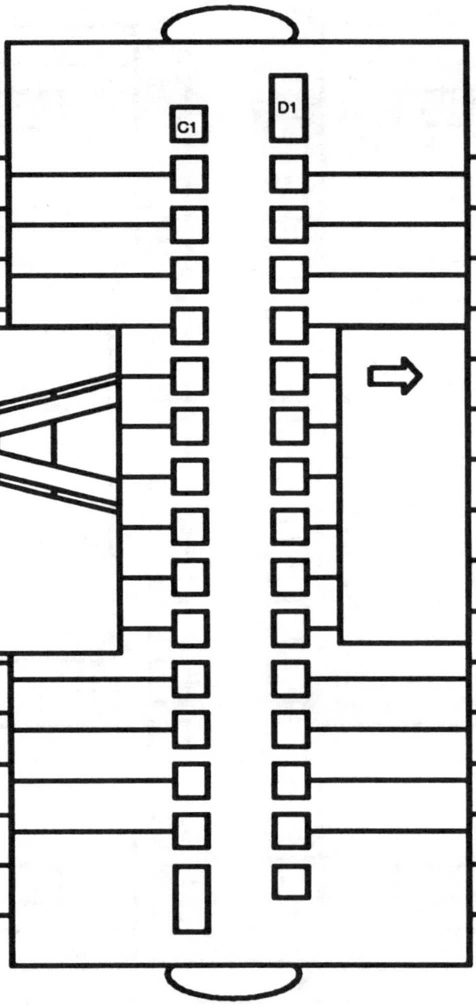

32 Pin C-D Connector

[3] Air Divert 4.3L & 2.8L

[4] Used for AC Rly 1.8L
Not used on most 4 cylinder

[5] Main Injector drive for most
4 cylinder

Pin Identifier

Type Three ECM Pin Connectors 24 and 32 Pins

Port Fuel Injection

Wire Color	Circuit I.D.	Pin
GRN/BLK	Fuel Pump	A1
[1]GRN/BLK	AC Cluch	A2
GRN/YEL	Can. Purge	A3
GRY	EGR Conrol	A4
BRN/WHT	CEL or SES	A5
PNK/BLK	Sw. Ign.	A6
TAN/BLK	TCC	A7
ORN	Serial Data	A8
WHT/BLK	Diagn.	A9
BRN	VSS	A10
[2]TAN	MAT 1.8L & 2.8L	A11
BLK/WHT	System Grd.	A12

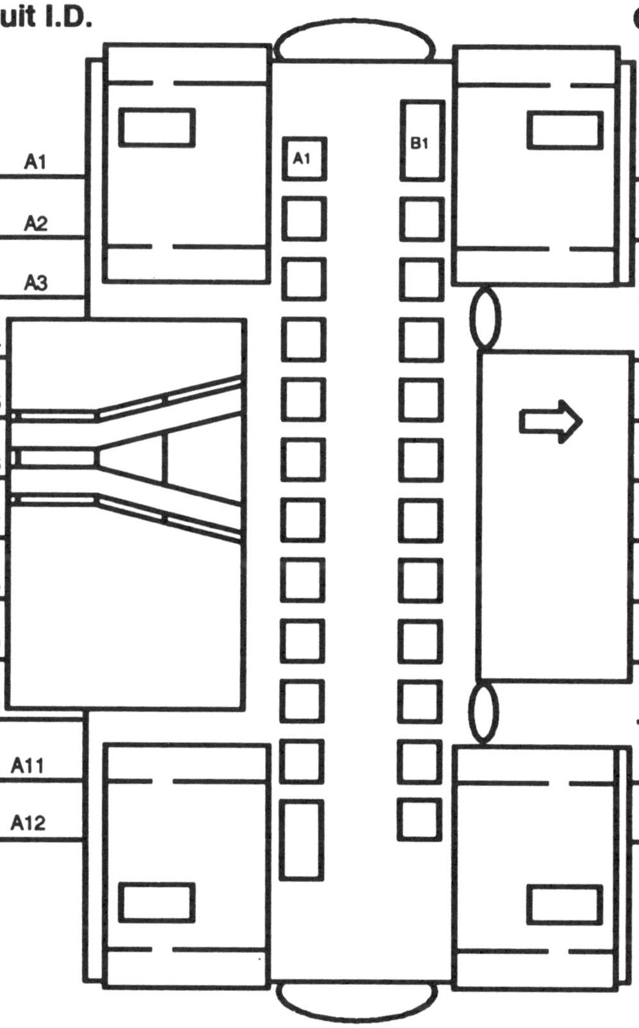

Pin	Circuit I.D.	Wire Color
B1	Battery	ORN
B2	Voltage Monitor	TAN/WHT
B3	HEI Ref. Grd.	BLK/RED
B4	EST Control	WHT
B5	HEI Ref.	PPL/WHT
B6	MAF 4&6 Cyl	BRN/WHT
B7	ESC	YEL/BLK
B8	AC Signal	DK GRN
B9	PS Switch	GRY
B10	P/N	ORN/BLK
B11	MAT 3.0L	
[3]B12	Inj. 5 3.8 Seq.	BLK/WHT

24 Pin A-B Connector

[1] Used for Air SW on 5.0L & 5.7L

[2] Used for Grd. on 1.8L

[3] Used for MAF on 5.0L & 5.7L

Pin Identifier

Type Three ECM Pin Connectors 24 and 32 Pins

Port Fuel Injection

Wire Color	Circuit I.D.	Pin
GRN/WHT	Cooling Fan	C1
[4]DK BRN	AC Relay	C2
LT GRN/BLK	IAC "B" Lo	C3
LT GRN/WHT	IAC "B" Hi	C4
LT BLU/WHT	IAC "A" Hi	
LT BLU/BLK	AC "B" Lo	6
[7]BLU	EGR Switch	C7
[8]Lt Blu	4th Gear Sw.	C8
[9]YEL	Cruise Control	C9
YEL	Coolant	C10
GRN	MAP	C11
BLK/YEL	Inj. #6 3.8L	C12
DK BLU	TPS	C13
GRY	5V Rev	C14
[10]LT GRN	Inj. Set A	C15
[11]ORN	Battery	C16

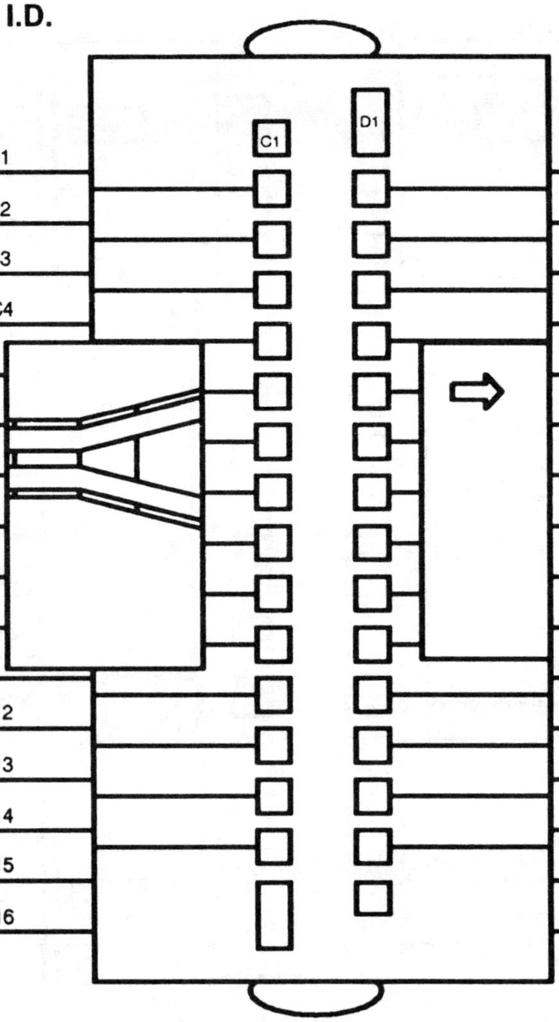

Pin	Circuit I.D.	Wire Color
D1	System Grd.	TAN/WHT
[5]D2	Cooling Fan	GRN/WHT
[6]D3	Cylinder Sol. Grd	BLK/WHT
D4	EST 1.8,5.0,5.7L	WHT
D5	5V Bypass	TAN/WHT
D6	0^2 Grd.	TAN
D7	0^2 Signal	PPL
D8	EGR Sw. 1.8 & 2.8L	WHT
D9	EGR Sw. 3.8L	PPL/WHT
D10	System Grd	BLK/WHT
D11	PS Sw. 3.0 & 3.8L	PPL
D12	CTS, MAT & TPS Grd.	TAN
D13	System Grd.	TAN
[12]D14	Inj. Set B	LT GRN
[13]D15	Inj. Set A	LT GRN
[14]D16	Inj. Set A	LT GRN

32 Pin C-D Connector

[4] Used for Air Divert on 5.0 & 5.7L
[7] Used for 3rd Gear Sw. on some Appl.
[10] Used for MAT or TPS backup on some Appl.
[13] Injector #3 on 3.8 Seq.

[5] Used as Coolant, MAT & TPS Grd on some engines
[8] PS Sw. on 1.8L
[11] Injector #1 on 3.8 Seq.
[14] Injector #4 on 3.8 Seq.

[6] Used for Wastegate control on some turbo applications
[9] Used as Crank on 1.8 & for AC Fan request on some.
[12] Injector #2 on 3.8 Seq, EGR Sw. on 5.0 & 5.7L

Pin Identifier

Type Four ECM Pin Connectors 24, 32 and 32 Pins

Throttle Body Fuel Injection

Black

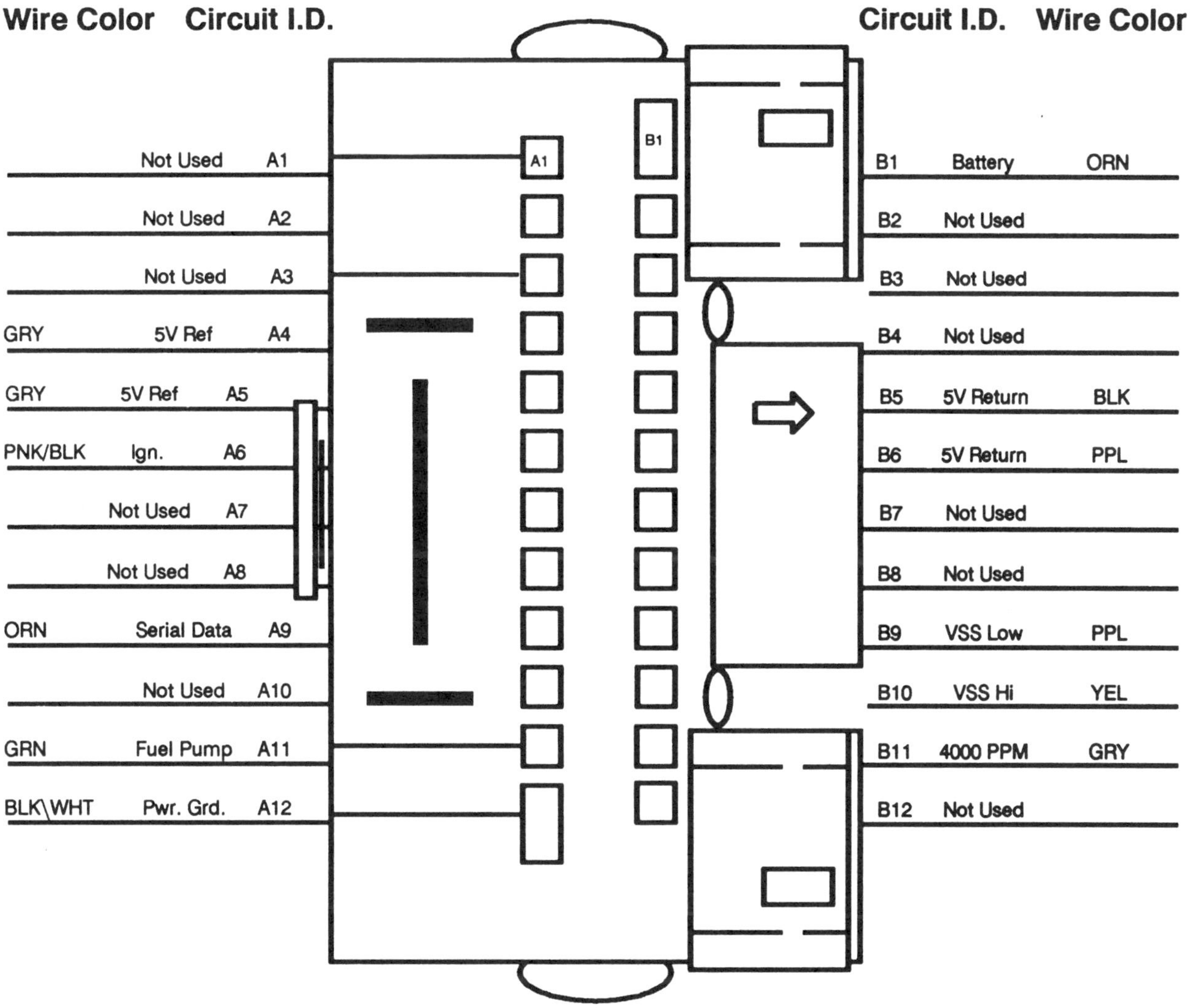

24 Pin A-B Connector

Pin Identifier

Type Four ECM Pin Connectors 24, 32 and 32 Pins

Throttle Body Fuel Injection

Black

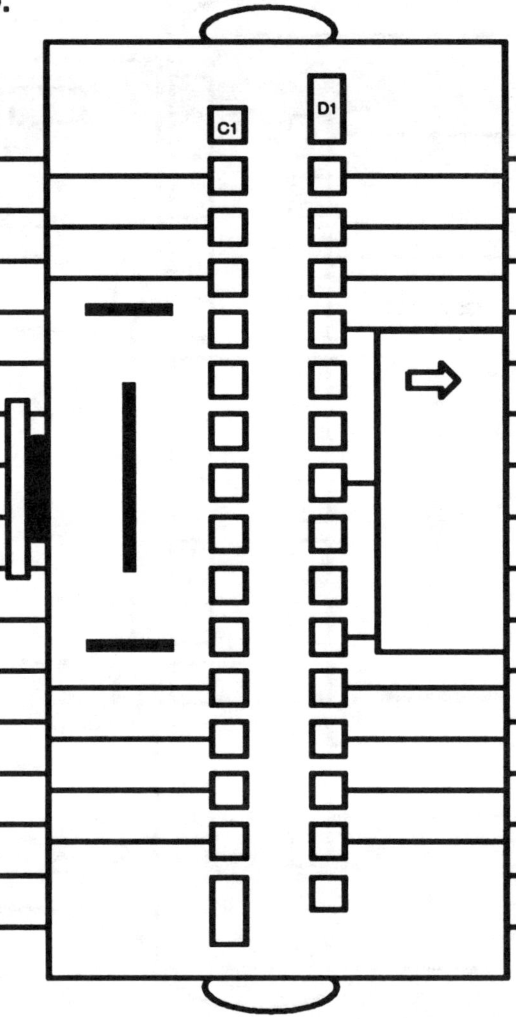

Wire Color	Circuit I.D.	Pin	Pin	Circuit I.D.	Wire Color
RED	2000 PPM	C1	D1	Power Grd.	
	Not Used	C2	D2	Not Used	
	Not Used	C3	D3	Not Used	
	Not Used	C4	D4	Not Used	
	Not Used	C5	D5	Jumper to D7	LT BLU
BRN	VSS	C6	D6	Not Used	
TAN/BLK	5V Bypass	C7	D7	Jumper to D5	LT BLU
WHT	EST	C8	D8	HEI Ref.	PPL/WHT
GRN/BLK	AC Signal	C9	D9	HEI Ref. Grd	BLK/RED
	Not Used	C10	D10	Not Used	
TAN	Injector Dr.	C11	D11	Not Used	
	Not Used	C12	D12	Not Used	
	Not Used	C13	D13	PS Sw.	BLK/WHT
	Not Used	C14	D14	Not Used	
	Not Used	C15	D15	Not Used	
ORN	Battery	C16	D16	Not Used	

32 Pin C-D Connector

Pin Identifier

Type Four ECM Pin Connectors 24, 32 and 32 Pins

Throttle Body Fuel Injection

Yellow

Wire Color	Circuit I.D.	Pin
	Not Used	E1
	Not Used	E2
LT BLU/WHT	IAC "A" Hi	E3
LT BLU/BLK	IAC "A" Lo	E4
LT GRN/WHT	IAC "B" Hi	E5
LT GRN/BLK	IAC "B" Lo	E6
BRN/WHT	CEL or SES	E7
GRN/WHT	Fan Rly	E8
	Not Used	E9
	Not Used	E10
	Not Used	E11
WHT/BLK	Diagnostic	E12
TAN/WHT	Fuel Pump	E13
PPL	0^2 Signal	E14
TAN	0^2 Grd.	E15
YEL	Coolant	E16

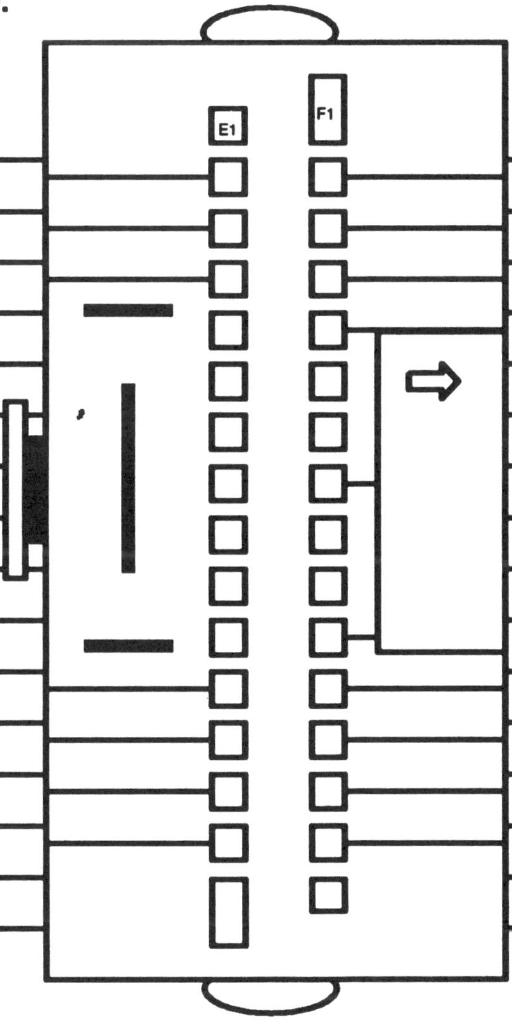

Pin	Circuit I.D.	Wire Color
F1	Not Used	
F2	Not Used	
F3	Not Used	
F4	AC Relay	BRN
F5	Not Used	
F6	TCC	TAN/BLK
F7	Not Used	
F8	Not Used	
F9	Not Used	
F10	Not Used	
F11	Not Used	
F12	Not Used	
F13	TPS	DK BLU
F14	Not Used	
F15	MAP	LT GRN
F16	MAT	TAN

32 Pin E-F Connector

Pin Identifier

Type Four ECM Pin Connectors 24, 32 and 32 Pins

Port Fuel Injection

Black

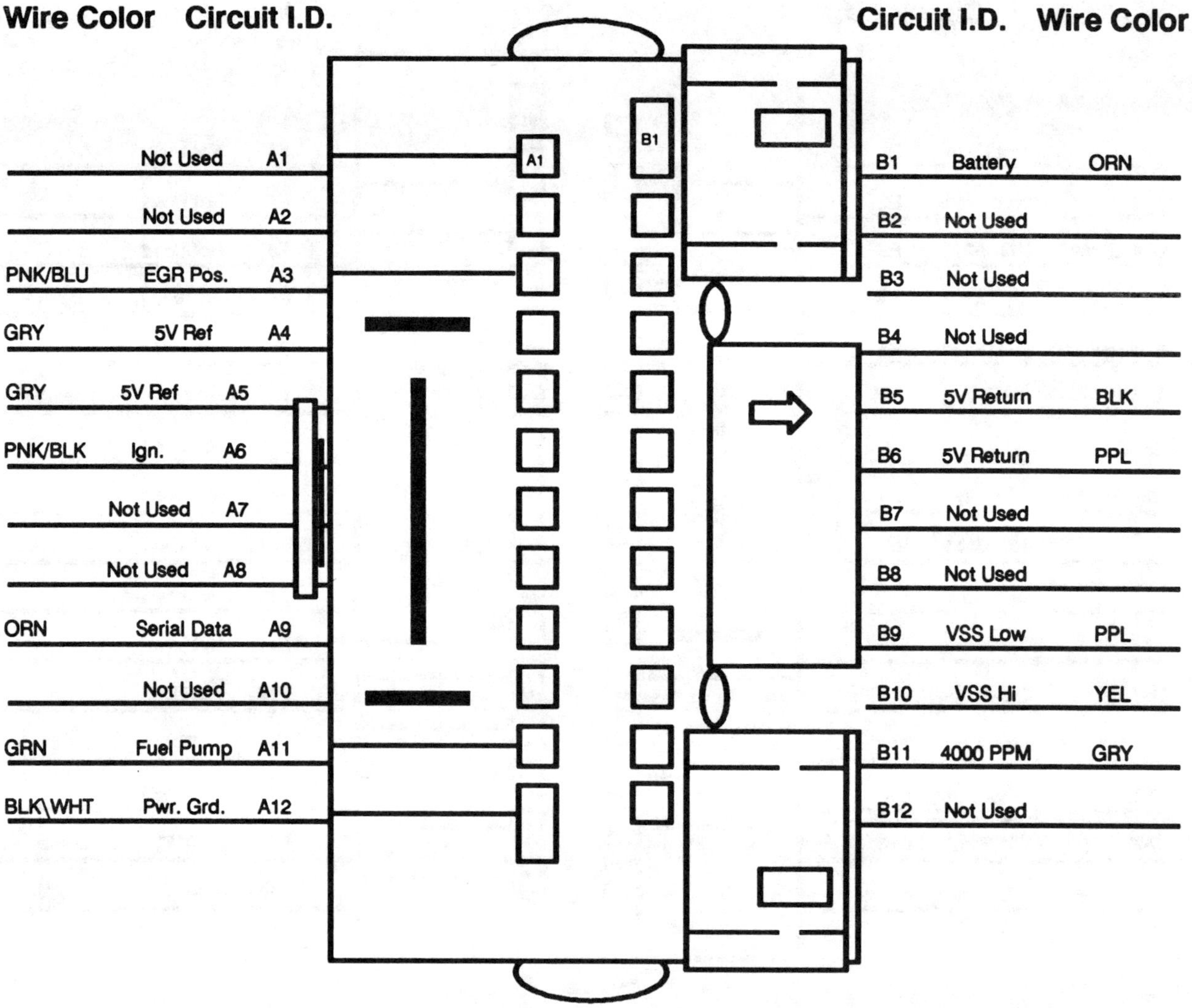

24 Pin A-B Connector

Pin Identifier

Type Four ECM Pin Connectors 24, 32 and 32 Pins

Port Fuel Injection

Black

Wire Color	Circuit I.D.	Pin
RED	2000 PPM	C1
	Not Used	C2
	Not Used	C3
	Not Used	C4
	Not Used	C5
BRN	VSS	C6
TAN/BLK	5V Bypass	C7
WHT	EST	C8
GRN/BLK	AC Signal	C9
	Not Used	C10
LT BLU	Injector Dr.1-3-5	C11
LT GRN	Injector Dr.2-4-6	C12
	Not Used	C13
	Not Used	C14
	Not Used	C15
ORN	Battery	C16

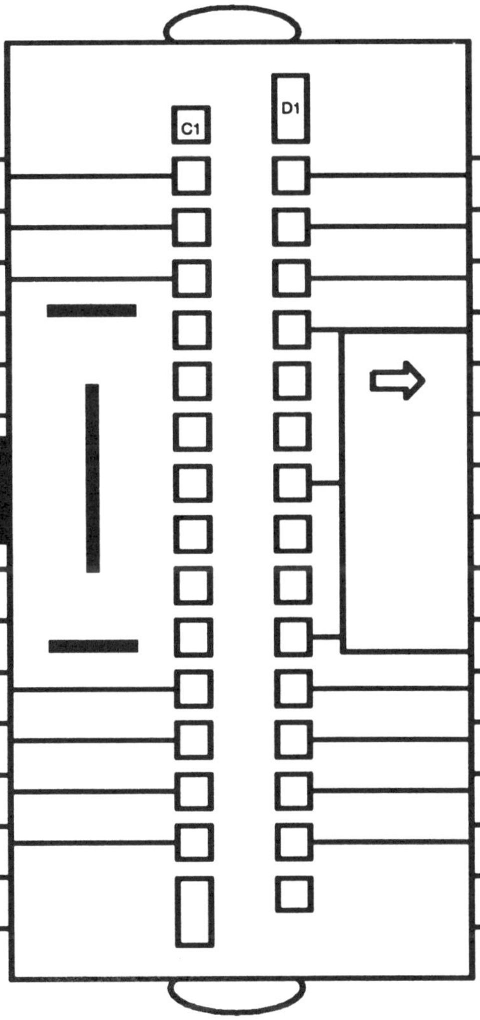

Pin	Circuit I.D.	Wire Color
D1	Power Grd.	
D2	Not Used	
D3	Not Used	
D4	Not Used	
D5	Inj. Dr Lo	BLK/WHT
D6	Inj. Dr Lo	BLK/WHT
D7	Jumper to D5	LT BLU
D8	HEI Ref.	PPL/WHT
D9	HEI Ref. Grd	BLK/RED
D10	Not Used	
D11	Not Used	
D12	AC Press. Fan	DK GRN/WHT
D13	PS Sw.	BLK/WHT
D14	4th Gear	LT BLU
D15	3rd Gear	DK GRN
D16	P/N	ORN/BLK

32 Pin C-D Connector

Pin Identifier

Type Four ECM Pin Connectors 24, 32 and 32 Pins

Port Fuel Injection

Yellow

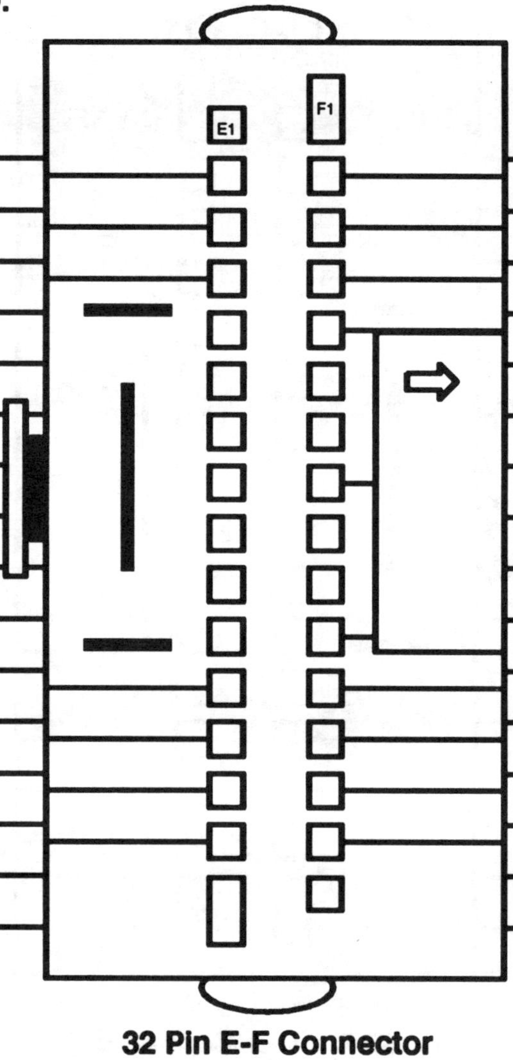

Wire Color	Circuit I.D.	Pin	Pin	Circuit I.D.	Wire Color
	Not Used	E1	F1	AC Rly	DK GRN/YEL
	Not Used	E2	F2	Not Used	
LT BLU/WHT	IAC "A" Hi	E3	F3	Not Used	
LT BLU/BLK	IAC "A" Lo	E4	F4	AC Relay	BRN
LT GRN/WHT	IAC "B" Hi	E5	F5	Not Used	
LT GRN/BLK	IAC "B" Lo	E6	F6	TCC	TAN/BLK
BRN/WHT	CEL or SES	E7	F7	CAN	DK GRN/YEL
GRN/WHT	Fan Rly	E8	F8	Not Used	
GRY	EGR	E9	F9	ESC	DK BLU
	Not Used	E10	F10	MAF	BRN/WHT
	Not Used	E11	F11	Not Used	
WHT/BLK	Diagnostic	E12	F12	Not Used	
TAN/WHT	Fuel Pump	E13	F13	TPS	DK BLU
PPL	0^2 Signal	E14	F14	Not Used	
TAN	0^2 Grd.	E15	F15	MAP	LT GRN
YEL	Coolant	E16	F16	MAT	TAN

32 Pin E-F Connector

Pin Identifier

Type Five ECM Pin Connectors 24, 32 and 32 Pins

Port Fuel Injection

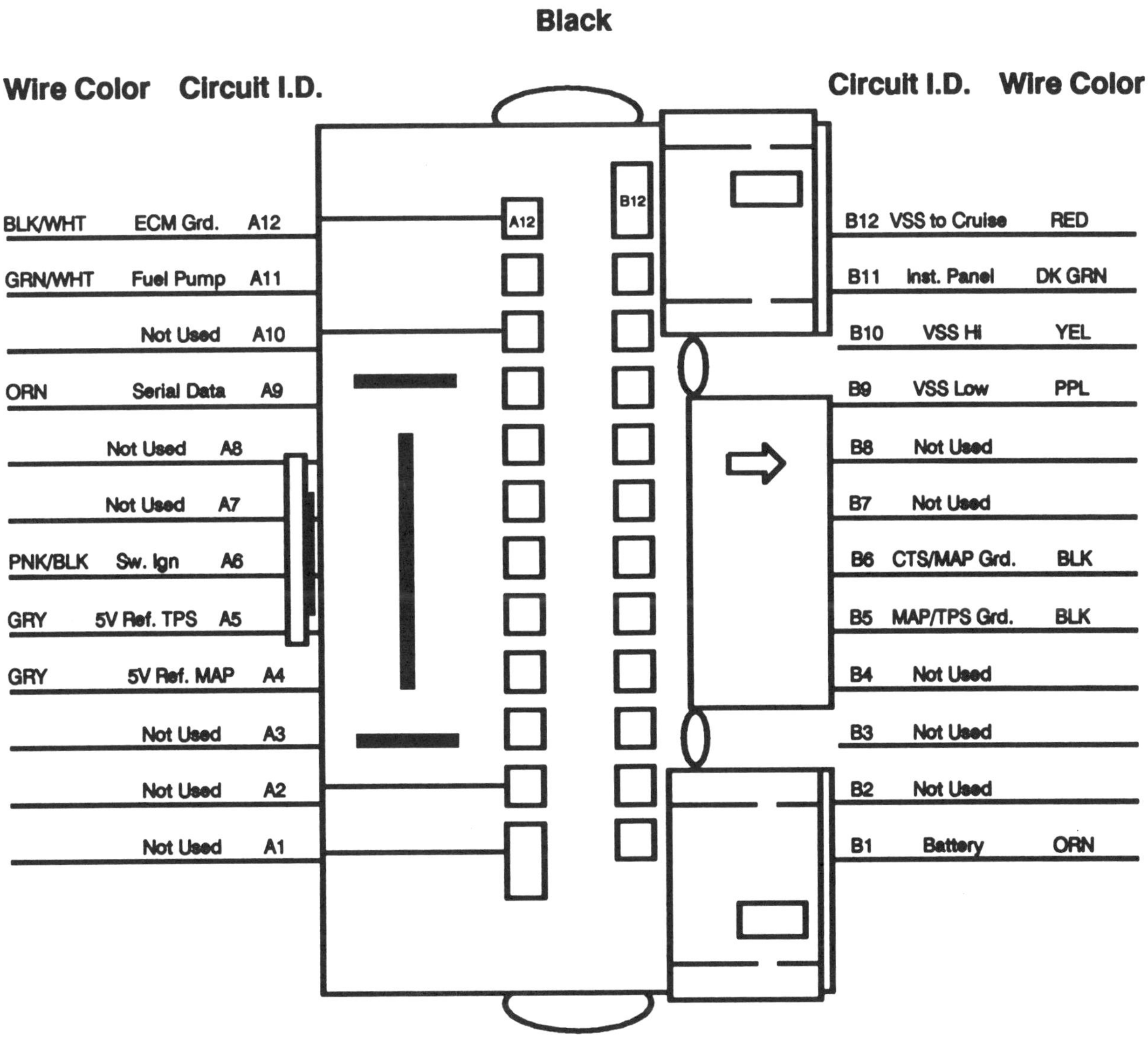

24 Pin A-B Connector

Pin Identifier

Type Five ECM Pin Connectors 24, 32 and 32 Pins

Port Fuel Injection

Black

Wire Color	Circuit I.D.	Pin
ORN	Battery	C16
PPL	Jumper to C13	C15
	Not Used	C14
PPL	Jumper to C15	C13
	Not Used	C12
	Not Used	C11
LT GRN	Inj. Dr.	C10
GRN/WHT	AC Rqst.	C9
WHT	EST	C8
TAN/BLK	5V Bypass	C7
BRN	VSS	C6
	Not Used	C5
	Not Used	C4
	Not Used	C3
	Not Used	C2
	Not Used	C1

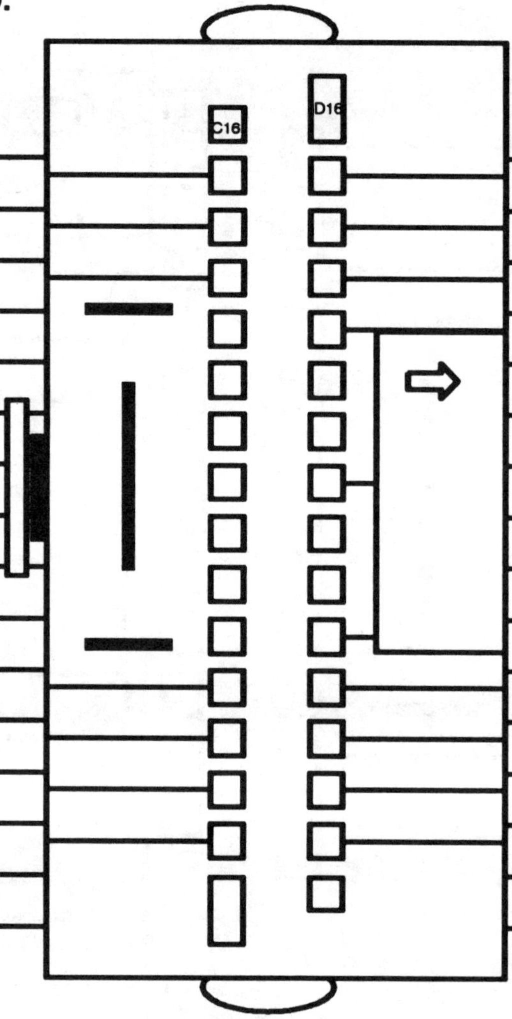

Pin	Circuit I.D.	Wire Color
D16	P/N	
D15	Not Used	
D14	EGR Sw.	WHIT
D13	PS Sw.	LT BLU/YEL
D12	Not Used	
D11	Not Used	
D10	Not Used	
D9	Ign. Grd.	BLK/RED
D8	HEI Ref	PPL/WHT
D7	Not Used	
D6	Inj. Jumper to D5	WHT
D5	AC Press. Fan	DK GR/WHT
D4	Not Used	
D3	Not Used	
D2	Not Used	
D1	ECM Grd.	BLK/WHT

32 Pin C-D Connector

Pin Identifier

Type Five ECM Pin Connectors 24, 32 and 32 Pins

Port Fuel Injection

Yellow

Wire Color	Circuit I.D.	Pin
YEL	Coolant	E16
TAN	0^2 Grd.	E15
PPL	0^2 Signal	E14
	Not Used	E13
WHT/BLU	Diagnostic	E12
	Not Used	E11
	Not Used	E10
BLK/GRN	EGR Control	E9
DK GRN/WHT	Fan Rly	E8
BRN/BLU	CEL or SES	E7
GRN/BLK	IAC "B" Lo	E6
GRN/WHT	IAC "B" Hi	E5
LT BLU/BLK	IAC "A" Lo	E4
LT BLU/WHT	IAC "A" Hi	E3
	Not Used	E2
	Not Used	E1

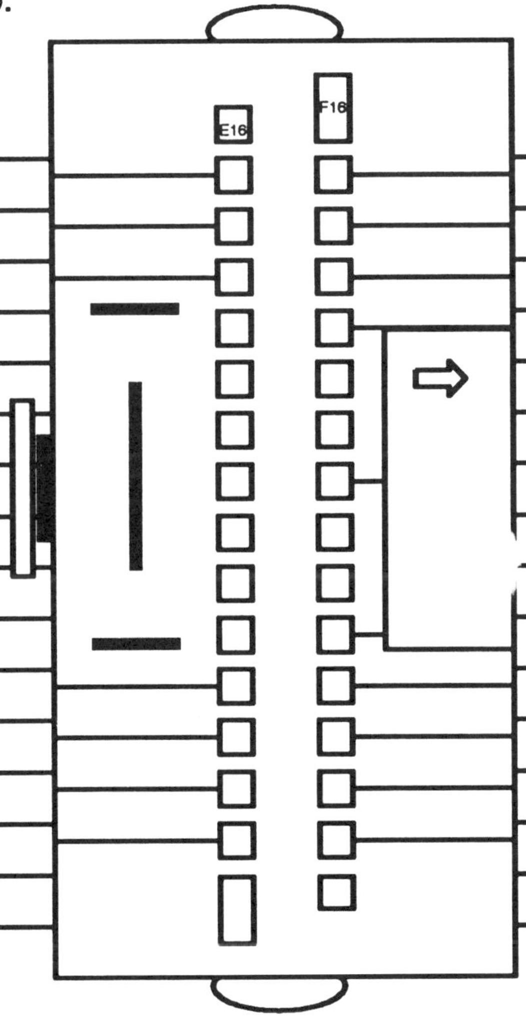

Pin	Circuit I.D.	Wire Color
F16	MAT	TAN
F15	MAP	LT GRN
F14	Not Used	
F13	TPS	DK BLU
F12	Not Used	
F11	Not Used	
F10	Not Used	
F9	ESC	DK BLU
F8	Not Used	
F7	Not Used	
F6	Not Used	
F5	Not Used	
F4	TCC	BRN/WHT
F3	Not Used	
F2	Wastegate Sol.	GRY
F1	AC Relay	DK BLU

32 Pin E-F Connector

Pin Identifier

Type Six ECM Pin Connectors 24, 32 and 32 Pins

Port Fuel Injection

Black

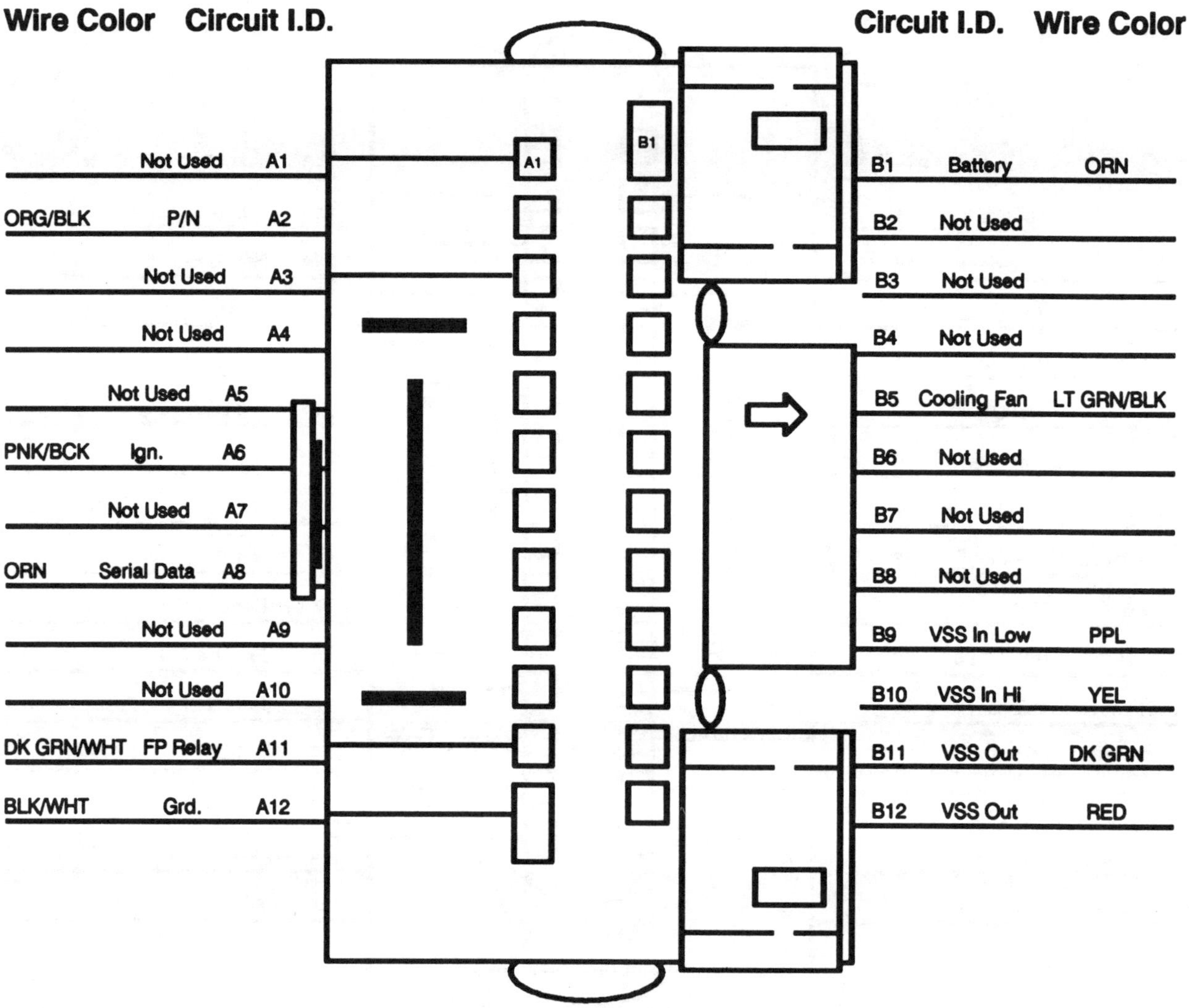

24 Pin A-B Connector

Pin Identifier

Type Six ECM Pin Connectors 24, 32 and 32 Pins

Port Fuel Injection

Black

Wire Color	Circuit I.D.	Pin
	Not Used	C1
RED	EGR #3	C2
BRN	EGR #2	C3
PPL	Brakes	C4
BLK	Cam Sig.	C5
	Not Used	C6
TAN/BLK	Bypass	C7
WHT	EST	C8
LT BLU	AC	C9
DK GRN	3rd Gear	C10
	Not Used	C11
	Not Used	C12
BLK/LT GRN	Inj. 2	C13
GRY	5V Ref.	C14
BLK/LT BLU	inj. 3	C15
ORN	Battery	C16

Pin	Circuit I.D.	Wire Color
D1	Grd	BLK/WHT
D2	EGR #1	BLU
D3	Not Used	
D4	PS	LT BLU/ORN
D5	Not Used	
D6	Grd	BLK/WHT
D7	Grd	BLK/WHT
D8	Ref. Hi	PPL/WHT
D9	Ref. Lo	BLK/RED
D10	Cruise	YEL
D11	Crank Ref	LT BLU/BLK
D12	Not Used	
D13	Inj. 4	BLK/RED
D14	Inj. 6	BLK/YEL
D15	Inj. 1	BLK
D16	Inj. 5	BLK/WHT

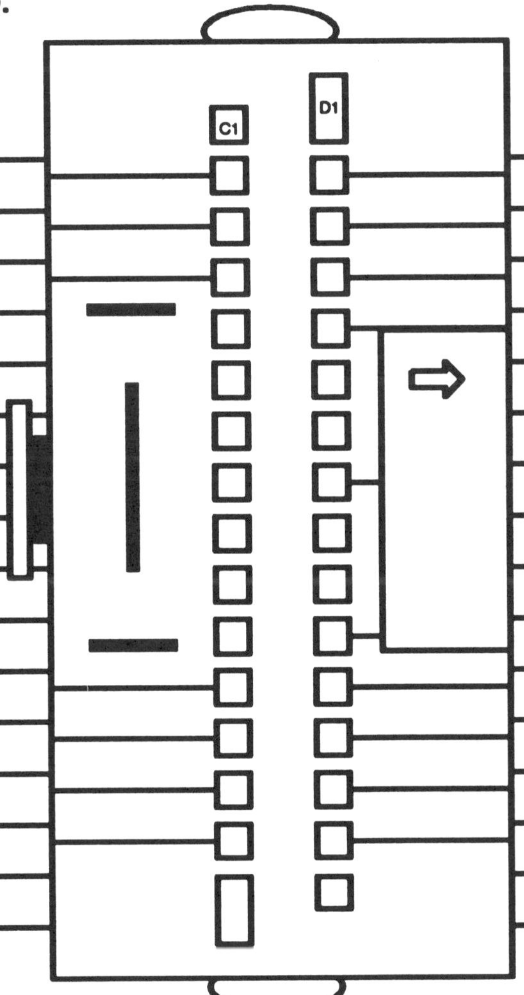

32 Pin C-D Connector

Pin Identifier

Type Six ECM Pin Connectors 24, 32 and 32 Pins

Port Fuel Injection

Yellow or Orange

Wire Color	Circuit I.D.	Pin
	Not Used	E1
DK GRN	Cooling Fan	E2
DK GRN/YEL	Can	E3
	Not Used	E4
LT BLU/WHT	IAC "A" Lo	E5
LT BLU/BLK	IAC "A" Hi	E6
LT GRN/WHT	IAC "B" Lo	E7
LT GRN/BLK	IAC "B" Hi	E8
DK GRN	CTS Sw.	E9
TAN/BLK	TCC	E10
BRN/WHT	SES	E11
	Not Used	E12
	Not Used	E13
	Not Used	E14
LT GRN	AC Comp	E15
	Not Used	E16

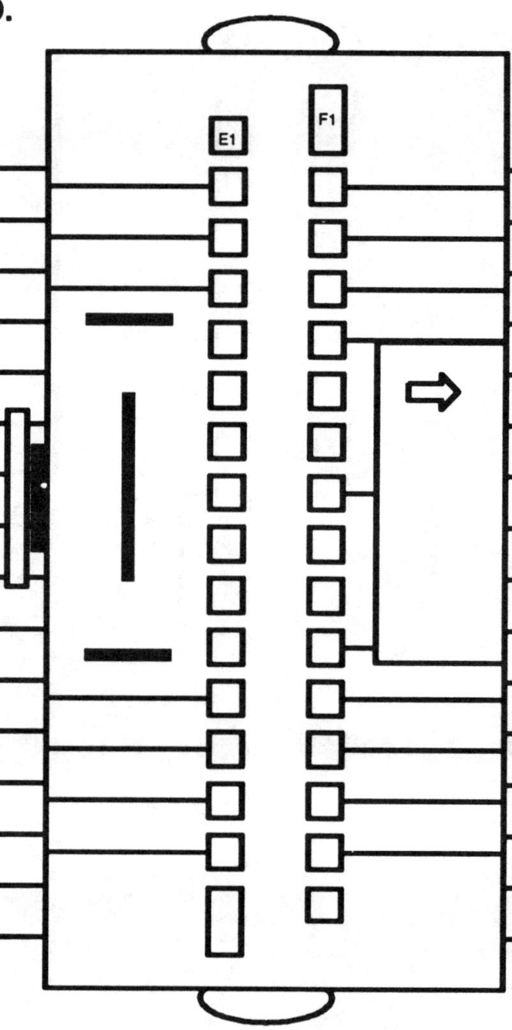

Pin	Circuit I.D.	Wire Color
F1	Not Used	
F2	0^2 Sig	BLK/PPL
F3	0^2 Ground	TAN
F4	CTS	YEL
F5	Signal Grd	BLK
F6	Diag	WHT/BLK
F7	4th Gear	LT BLU
F8	2nd Gear	ORN
F9	ESC	DK BLU
F10	MAF	YEL
F11	MAT	TAN
F12	Not Used	
F13	TPS	DK BLU
F14	Not Used	
F15	Not Used	
F16	Not Used	

32 Pin E-F Connector

Using The Specific Trouble Code Lists

It would be wonderful if all the trouble codes had only one possible definition, but that's not the case. Many diagnostic errors have been made in the field because of a misdefined trouble code. The fact of the matter is that the trouble code definition on each engine platform truly only has one definition. The variations arise because of the large number of different engine platforms and the inputs and outputs used for each one.

For this reason the Logical Sequence Diagnosis requires the technician to carefully identify the vehicle, the system used and the devices used with that system. The General Fault Code Definitions are listed in Section III following the discussion on how to enter the On-Board Diagnostic System and retrieve fault codes. In many cases the proper definition will be obvious, in others the proper definition will depend on which year model is being tested and which device is used on the system in question. The following Specific Fault Code Definitions are a more detailed listing of the possible fault code definitions broken down by year model. Even when broken down, there will still be some fault codes with multiple definitions. That is where pin-pointing the specific definition will require the identification of the device or circuit is being used for the vehicle displaying the fault code.

In summary, after retrieving any fault codes, refer first to the General Fault Code Definitions. If this sufficiently defines the code for further diagnosis, you won't need to use the Specific Fault Code Definitions. If the code has multiple definitions, use the Specific Fault Code Definitions and see if defining by year model will narrow it down. If not, then carefully identify the ECM circuits and/or the devices used to pin-point the specific definition so that you can proceed with the correct troubleshooting steps.

Specific Trouble Codes

1982

Code	Definition
12	No distributor reference pulse
13	Oxygen sensor circuit
14	Coolant sensor
15	Coolant sensor
21	Throttle position high
22	Throttle position low
23	M/C circuit solenoid low
24	Vehicle speed sensor
31	Wastegate solenoid
32	BARO sensor
33	MAP sensor high
34	Differential pressure (vacuum)
	Or
34	MAP sensor circuit
35	Idle speed control circuit
41	No distributor reference signal
42	Open/shorted bypass or EST

Specific Trouble Codes

1982 (Cont.)

Code	Definition
43	Electronic spark control
44	Lean exhaust indication
45	Rich exhaust indication
44/45	At same time, faulty oxygen sensor
51	Faulty PROM or improper installation
54	Shorted M/C solenoid or faulty ECM
55	Ground voltage reference, faulty oxygen sensor or faulty ECM

Specific Trouble Codes

1983

Code	Definition
12	No distributor reference pulse
13	Oxygen sensor circuit
14	Coolant sensor
15	Coolant sensor
17	Transmission Lock-Up Solenoid
21	Throttle position high
22	Throttle position low
23	M/C circuit solenoid low
24	Vehicle speed sensor
31	Wastegate solenoid
32	BARO sensor
33	MAP sensor
34	Differential pressure (vacuum)
35	Idle speed control circuit
41	No distributor reference signal
42	Open/shorted bypass or EST
43	Electronic spark control
44	Lean exhaust indication
45	Rich exhaust indication
51	Faulty PROM or improper installation
54	Shorted M/C solenoid or faulty ECM
55	Ground VREF, faulty oxygen sensor or faylty ECM

Specific Trouble Codes

1984

Code	Definition
12	No distributor reference pulse
13	Oxygen sensor circuit
14	Coolant sensor
15	Coolant sensor
21	Throttle position high
	Or
21	Idle and WOT switch error
22	Throttle position low
	Or
22	Fuel cutoff solenoid low
23	M/C circuit solenoid low
	Or
23	MAT sensor low
24	Vehicle speed sensor
25	MAT sensor high
	Or
25	Air switch solenoid low
31	Wastegate electrical signal open
32	EGR circuit malfunction

Specific Trouble Codes

1984 (Cont.)

Code	Definition
33	Mass flow sensor
34	Differential pressure (vacuum)
	Or
34	MAP sensor circuit
	Or
34	Mass flow sensor (low/no signal)
35	Idle speed control
41	No distributor reference signal
	Or
41	Cam sensor failed
42	Electronic spark control
	Or
42	Error in Distributor or C3 system
	Or
42	Fuel cutoff solenoid high
43	Electronic spark control
44	Lean exhaust indication
45	Rich exhaust indication
51	Faulty PROM or improper installation

Specific Trouble Codes

1984 (Cont.)

Code	Definition
52	Calpack missing
53	EGR vacuum control
	Or
53	Air switch solenoid high
54	Shorted M/C solenoid or faulty ECM
55	Ground voltage reference, faulty oxygen sensor or faulty ECM

Specific Trouble Codes

1985

Code	Definition
12	No distributor reference pulse
	Or
12	Engine speed active
13	Oxygen sensor circuit
14	Coolant sensor high
15	Coolant sensor low
21	Throttle position high
22	Throttle position low
23	M/C circuit solenoid low
	Or
23	MAT sensor low
	Or
23	Air fuel solenoid
24	Vehicle speed sensor
25	MAT sensor high
31	Wastegate electrical signal open
	Or
31	Wastegate high MAP overboost
	Or

Specific Trouble Codes

1985 (Cont.)

Code	Definition
31	MAP low
32	BARO sensor
	Or
32	EVRV/EGR
33	Mass airflow sensor (high)
	Or
33	Incompatible PROM
34	MAP sensor low
	Or
34	Differential pressure (vacuum)
	Or
34	Mass airflow sensor (low)
35	Idle Air Control (IAC) error
	Or
35	ISC nose switch shorted
41	No distributor reference signal
	Or
41	Cam sensor failed
	Or

Specific Trouble Codes

1985 (Cont)

Code	Definition
41	Cylinder select error
42	EST monitor error
	Or
42	Bypass line error
43	Electronic spark control
44	Lean exhaust indication
45	Rich exhaust indication
51	Fault PROM or improper installation
	Or
51	Check sum error
	Or
51	PROM
52	Calpack missing
	Or
52	A/D ECM failure
53	Over voltage
	Or
53	Water injection failure
	Or

Specific Trouble Codes

1985 (Cont)

Code	Definition
53	Voltage Reference (VREF)
54	M/C solenoid high
	Or
54	Fuel pump voltage
	Or
54	Fuel pump relay malfunction
55	Faulty oxygen sensor or faulty ECM

Specific Trouble Codes

1986

Code	Definition
12	No distributor reference pulse
13	Oxygen sensor circuit
14	Coolant sensor high
15	Coolant sensor
	Or
15	Coolant switch
21	Throttle position sensor
	Or
21	Throttle position high
22	Throttle position high
23	M/C circuit solenoid low
	Or
23	MAT sensor low
	Or
23	Air fuel solenoid
24	Vehicle speed sensor
25	MAT sensor high
31	Wastegate electrical signal open
	Or

Specific Trouble Codes

1986 (Cont.)

Code	Definition
31	Wastegate high MAP overboost
32	ER diagnostics
	Or
32	BARO sensor
	Or
32	EVRV/EGR
33	Mass airflow sensor
	Or
33	MAP sensor high
34	MAP sensor low
	Or
34	Differential pressure (vacuum)
	Or
34	Mass airflow sensor (low)
35	Idle Air Control
	Or
35	ISC nose switch shorted
	Or
35	IAC diagnostics

Specific Trouble Codes

1986 (Cont.)

Code	Definition
36	Burnoff diagnostics
41	No distributor reference signal
	Or
41	Cam sensor failed
	Or
41	Cylinder select error
42	EST monitor error
	Or
42	Bypass line error
43	Electronic spark control
44	Lean exhaust indication
45	Rich exhaust indication
51	Fault PROM
52	EGR failure on V8
53	High battery voltage diagnostic
	Or
53	Over voltage
	Or
53	Water injection failure

Specific Trouble Codes

1986 (Cont.)

Code	Definition
51	Faulty PROM or improper installation
52	Calpack missing
53	EGR failure on V8
	Or
53	Over voltage
	Or
53	Water injection failure
54	M/C solenoid high
	Or
54	Fuel pump voltage
	Or
54	Fuel pump relay malfunction
55	Grounded voltage reference, faulty oxygen sensor or faulty ECM

Specific Trouble Codes

1987

Code	Definition
12	No distributor pulse
13	Oxygen sensor
14	Coolant sensor high temperature
15	Coolant sensor low temperature
	Or
15	Coolant switch
21	TPS/idle switch fault
	Or
21	Throttle position sensor high
22	Throttle position sensor low
23	M/C solenoid
	Or
23	MAT sensor low/open
24	Vehicle speed sensor
25	MAT sensor high/shorted
31	Wastegate overboost
	Or
31	Wastegate pulse not detected

Specific Trouble Codes

1987 (Cont.)

Code	Definition
32	EGR diagnostics
	Or
32	BARO sensor
33	Mass airflow sensor high
	Or
33	MAP sensor high
34	MAP sensor low
	Or
34	Mass airflow sensor low
	Or
34	MAP sensor
35	IAC error
36	Burnoff diagnostics
41	Cylinder select error
	Or
41	Tach input error
	Or
41	Cam sensor
42	EST error

Specific Trouble Codes

1987 (Cont.)

Code	Definition
42	Bypass error
43	ESC failure or direct fire in bypass
44	Oxygen sensor lean
45	Oxygen sensor rich
46	VATS failure
51	PROM/memory error
	Or
51	Calibration PROM
52	Calpack missing
53	EGR
	Or
53	High battery voltage
	Or
53	Water injection failure
54	Fuel pump relay/voltage
	Or
54	Idle mixture (CO) requires adjustment
	Or
54	M/C solenoid

Specific Trouble Codes

1987 (Cont.)

Code	Definition
55	ECM failure (ADU error)
56	Corrosivity/Add coolant
61	Oxygen sensor degraded
63	MAP sensor high
64	MAP sensor low

Specific Trouble Codes

1988

Code	Definition
12	No distributor reference pulse
13	Oxygen sensor
14	Coolant sensor high temperature
15	Coolant sensor low temperature
16	High battery voltage
21	TPS/idle switch fault
	Or
21	Throttle position hsensor high
22	Throttle position low
	Or
22	Fuel cutoff relay circuit open/grounded
23	M/C solenoid
	Or
23	MAT sensor low/open
24	Vehicle speed sensor
25	MAT sensor high/shorted
	Or
25	Vacuum switching valve circuit open/shorted
26	Quad Driver Module (QDM)

Specific Trouble Codes

1988 (Cont.)

Code	Definition
27	2nd gear switch
28	3rd gear switch
29	4th gear switch
31	Wastegate overboost
	Or
31	Wastegate pulse not detected
	Or
31	Fuel injector
	Or
31	Park/neutral switch
32	EGR diagnostics
33	BARO sensor
34	MAP sensor flow
	Or
34	Mass airflow sensor low
	Or
34	MAP sensor
35	IAC error
	Or

Specific Trouble Codes

1988 (Cont.)

Code	Definition
35	ISC nose switch
36	Burnoff diagnostics
	Or
36	Vacuum leak
38	Brake switch
39	TCC failure
41	Cylinder select error
	Or
41	Tach input error
	Or
41	Cam sensor
42	EST error
	Or
42	Bypass line error
	Or
42	Fuel cutoff relay/circuit shorted
	Or
42	EST bypass low
43	ESC failure or direct fire in bypass

Specific Trouble Codes

1988 (Cont.)

Code	Definition
44	Oxygen sensor lean
45	Oxygen sensor rich
46	VATS failure
	Or
46	Power steering switch
47	UART link
48	Misfire
49	Vacuum leak
51	PROM/memory error
52	Calpack missing
53	EGR
	Or
53	High battery voltage
53	Water injection failure
	Or
53	Vehicle Anti-Theft System (VATS) diagnostics
54	Fuel pump relay/voltage
	Or

Specific Trouble Codes

1988 (Cont.)

Code	Definition
54	M/C solenoid
55	ECM failure (ADU error)
56	Corrosivity/Add coolant
58	Vehicle Anti-Theft System (VATS) problem
61	Oxygen sensor degraded
62	Gear switch inputs diagnostics
63	Small EGR failure
	Or
63	MAP sensor high
64	MAP sensor low
	Or
64	Medium EGR failure
65	Large EGR failure
	Or
65	Injector peak/hold diagnostic
66	A/C pressure transducer fault
	Or
66	ECM reset

Specific Trouble Codes

1989

Code	Definition
12	No distributor reference pulse
13	Oxygen sensor
14	Coolant sensor high temperature
15	Coolant sensor low temperature
16	High battery voltage
21	Throttle position sensor high
22	Throttle position sensor low
23	MAT sensor low/open
	Or
23	TPS error
24	Vehicle speed sensor
25	MAT sensor high/shorted
26	Quad Driver Module (QDM)
27	2nd gear switch
28	3rd gear switch
29	4th gear switch
31	Wastegate overboost
	Or
31	MAP low

Specific Trouble Codes

1989 (Cont.)

Code	Definition
	Or
31	Fuel injector
	Or
31	Park/Neutral switch
32	EGR diagnostics
34	MAP sensor low
	Or
34	MAF sensor low
35	Improper idle
36	Burnoff diagnostics
38	Brake switch
39	TCC failure
41	Missing 1x
	Or
41	Cam sensor
42	EST error
	Or
42	Bypass line error
	Or

Specific Trouble Codes

1989 (Cont.)

Code	Definition
42	EST bypass low
43	EQU code 42A
44	Oxygen sensor lean
45	Oxygen sensor rich
46	VATS failure
	Or
46	Power steering switch
47	UART link
48	Misfire
49	Vacuum leak
51	PROM/memory error
	Or
51	Check sum error
52	Calpack missing
	Or
52	QDM fault
	Or
52	A/D error
53	High battery voltage

Specific Trouble Codes

1989 (Cont.)

Code	Definition
	Or
53	V reference error
	Or
53	Vehicle anti-theft system (VATS) diagnostics
	Or
53	Vacuum switching valve voltage high
54	Fuel pump relay/voltage
55	ECM failure (ADU error)
	Or
55	Serial bus error
	Or
55	SAD error
56	Corrosivity/add coolant
58	VATS problem
61	Oxygen sensor degraded
62	Gear switch inputs diagnostics
63	Small EGR failure
64	Medium EGR failure

Specific Trouble Codes

1989 (Cont.)

Code	Definition
65	Large EGR failure
	Or
65	Injector peak/hold diagnostic
66	A/C pressure transducer fault
	Or
66	ECM reset
88	ECM reset

Specific Trouble Codes

1990

Code	Definition
12	No distributor reference pulse
13	Oxygen sensor
14	Coolant sensor high temperature
15	Coolant sensor low temperature
16	High battery voltage
	Or
16	DIS fault line open/shorted
17	RPM signal problem
21	Throttle position sensor high
22	Throttle position sensor low
23	MAT sensor low/open
	Or
23	TPS error
	Or
23	ATS sensor low
24	Vehicle speed sensor
25	MAT sensor high/shorted
	Or
25	ATS sensor high

Specific Trouble Codes

1990 (Cont.)

Code	Definition
26	Quad Driver Module (QDM)
27	2nd gear switch
28	3rd gear switch
29	4th gear switch
31	Wastegate overboost
	Or
31	MAP low
	Or
31	Fuel injector
	Or
31	Park/Neutral switch
	Or
31	Governor malfunction
	Or
31	Cam diagnostic
32	EGR diagnostics
34	MAP sensor low
	Or
34	MAF sensor low

Specific Trouble Codes

1990 (Cont.)

Code	Definition
35	IAC error
36	Transmission shift problem
	Or
36	DIS fault indicating extra/miss EST
38	Brake switch
39	TCC failure
41	Cylinder select error
	Or
41	Missing 1X
	Or
41	Cam sensor
42	EST error
	Or
42	EST bypass low
43	ESC failure or direct fire in bypass
	Or
43	EQU code 42A
44	Oxygen sensor lean
45	Oxygen sensor rich

Specific Trouble Codes

1990 (Cont.)

Code	Definition
46	VATS failure
	Or
46	Power steering switch
47	UART link
48	Misfire
49	Vacuum leak
51	PROM/memory error
	Or
51	Check sum error
52	Calpack missing
	Or
52	A/D error
	Or
52	Oil temperature sensor low
53	High battery voltage
	Or
53	V reference error
	Or
53	Vehicle anti-theft system (VATS) diagnostics

Specific Trouble Codes

1990 (Cont.)

Code	Definition
	Or
53	Governor malfunction
54	Fuel pump relay/voltage
	Or
54	QDM component failure
55	ECM failure (ADU error)
	Or
55	Fuel lean malfunction
56	Corrosivity/add coolant
	Or
56	Port throttle system vacuum sensor
58	VATS problem
61	Oxygen sensor degraded
	Or
61	Port throttle system error
	Or
61	Cruise vent solenoid circuit
62	Gear switch inputs diagnostics
	Or

Specific Trouble Codes

1990 (Cont.)

Code	Definition
62	Oil temperature sensor high
	Or
62	Cruise vacuum solenoid circuit
63	Small EGR failure
	Or
63	Right oxygen sensor failed
64	Right oxygen sensor lean
	Or
64	Medium EGR failure
65	Large EGR failure
	Or
65	Injector peak/hold diagnostic
	Or
65	Right oxygen sensor rich
	Or
65	Cruise servo position sensor
66	A/C pressure transducer fault
	Or
66	ECM reset

Specific Trouble Codes

1990 (Cont.)

Code	Definition
67	Cruise switches circuit
68	Cruise system circuit
69	A/C pressure switch circuit
88	ECM reset

Review of Key Concepts

- **Using the Pin Identifiers allows the technician to test the circuits at the ECM.**

- **All sensor inputs terminate at the ECM.**

- **All sensor references originate at the ECM.**

- **All actuator controls originate at the ECM.**

- **Testing an ECM pin for voltage will save the technician a lot of time for circuit troubleshooting.**

Table of Contents

Section V

Section V Outline

V. Common GM EFI Vehicle Tests & Technical Service Bulletins

A. A complete EFI diagnosis always contains a fuel delivery test.
   1. Test the system for pressure.
   2. Test the system for good fuel flow.
   3. Test the vehicle fuel for any quality problems.

B. Injector cleaning is the tune–up of the EFI engines.
   1. To accomplish the task, make the proper hook–up
   2. Make sure the cleaning solution enters the fuel injectors and not the fuel tank.

C. Be prepared to check and adjust the minimum idle speed.

D. An injector balance test will verify if the individual injectors are delivering fuel at the same rate.

E. The Technical Service Bulletins are a great aid in diagnosing GM EFI problems.

Fuel Management System Tests

When GM began manufacturing computer controlled carbureted engines, GM engineers recognized the need for a comprehensive troubleshooting procedure to verify the system's fuel management capability. The troubleshooting "trees" contained a very simple procedure called the "SYSTEMS PERFORMANCE CHECK". For this procedure the technician uses a simple dwell-tach and jumper lead to verify that the carburetor is capable of changing the A/F ratio and to check the carburetor calibration. The subsequent test steps and additional troubleshooting trees can then be used to verify the ECM's capability to issue the control commands to facilitate good fuel management.

For EFI engines, a simple dwell meter cannot be used for monitoring ECM fuel control commands since the injector control pulses are issued at varying pulse rates. The data stream on some GM EFI vehicles contains a word for INJECTOR PULSE WIDTH, but that alone is not a very conclusive piece of information since the variables affecting the injector pulse width are so numerous.

However, there are other more informative indicators which can be used to verify the ECM's fuel management capability. After verifying that the fuel delivery system is delivering adequate fuel pressure and flow, the technician can monitor the exhaust oxygen content to establish the overall fuel management capability of the system. Be sure and test the O^2 content for each engine operating mode. Test the O^2 at idle, cruise and under a load. Road test the vehicle and observe the O^2 at steady cruise and during acceleration and deceleration. If the O^2 voltage is switching back and forth above and below the Lambda switch point (approximately .450V) when the engine is idled, cruising or under no load, then the ECM can be said to be able to control the fuel commands within acceptable fuel management ranges. When the vehicle is accelerated, the O^2 voltage should rapidly increase to above .750V. When the vehicle is decelerated, the O^2 voltage should decrease to below .100V. Readings of this kind verify that the ECM is capable of sending additional fuel for acceleration enrichment and is also capable of reducing the fuel for deceleration enleanment. Monitoring the O^2 voltage under different operating conditions can then establish the overall system fuel management capability.

In addition to the O^2 voltage trends, the key fuel management input signal indicator, the technician can also observe the output command called "Integrator". The Integrator is a numerical factor which the ECM applies to its fuel injector pulse width commands to compensate for various engine fuel requirement conditions. For instance, as the engine ages or deteriorates, the ECM has the ability to factor additional fuel into the fuel commands to compensate for the engine deterioration. The ECM can also use the Integrator to add or subtract fuel from the calculated pulse commands as a result of O^2 voltage trends which have indicated to the ECM

Fuel Management System Tests

Cont.

that the vehicle operator requires additional fuel or less fuel. Some manufacturers use the term "ADAPTIVE" to describe this ability. The ECM establishes fuel factoring adjustments from the Integrator and stores then for future vehicle operation in the "BLOCK LEARN" cells. GM ECMs can have up to 16 different Block Learn cells. Each cell represents a different engine operating condition based on throttle position, RPM and engine load. The Block Learn fuel factoring adjustments are then called upon by the ECM to adapt or modify the fuel commands throughout subsequent vehicle operation. The numerical representation for the Integrator and Block Learn factors are based on the number 256. Theoretically, the ECM could adapt the fuel commands by as much as half of 256 lean or half of 256 rich. The reason the number 256 is used relates to the design of the microprocesor in the ECM. The normal Block Learn or Integrator factor numbers will not exceed a certain window or range such as, no lower than 110 nor higher than 150. The illustration below demonstrates the relationship of Block Learn and Integrator to engine fuel commands from the ECM.

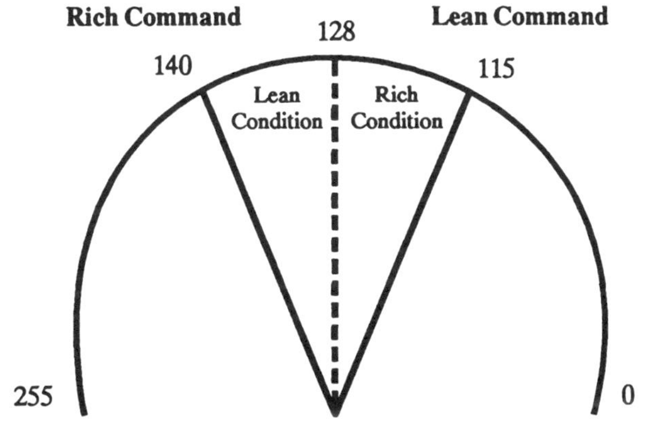

Remember, the injector pulse width is the length of time the injector(s) remain open, delivering fuel into the engine. The Block Learn/Integrator numbers represent the factor which the ECM is applying to pulse width calculations for fuel requirement adjustment.

Therefore, if the O^2 voltage is switching above and below Lambda, and the Block Learn/Integrator is at or near 128, then the ECM is managing fuel very well, right in the middle of the systems' fuel control range. If the O^2 voltage is switching, but the Block Learn/Integrator counts are way above or below the 128, then the ECM is still managing fuel, but is having to factor the fuel commands to accomplish fuel management. When the O^2 voltage is no longer switching, or the Block Learn cells or Integrator are stuck at either end of the operating range, then the ECM has run out of rope.

Fuel Management System Tests

Cont.

Monitoring these input and output signals and emulating changes in these signals allows the technician to more fully analyze the ECM's fuel management capability. Here's a simple procedure for testing the system. You could call it a SYSTEMS PERFORMANCE CHECK for EFI engines.

Monitor O^2 and Block Learn/Integrator at various engine Air/Fuel Requirements

Emulate Full Rich
- High CTS Voltage
- High MAT Voltage
- High MAP Voltage
- Low VAC Voltage
- High MAF Frequency
- Restrict the Air or
- Increase the Fuel

Note: If the ECM responds inversely to the inputs, then it is "seeing" the inputs and sending the correct commands.

Monitor for:
- Decreasing O^2 Voltage
- Decreasing Block Learn/Integrator
- Decreasing Injector Pulse Width

Emulate Full Lean
- Low MAP Voltage or
- High VAC Voltage
- Low MAF Frequency
- Restrict the Fuel or
- Increase the Air

Monitor for:
- Increasing O^2 Voltage
- Increasing Block Learn/Integrator
- Increasing Injector Pulse Width

Watch out for:
- AIR management system problems
- Canister Purge system problems
- Engine oil contaminated with fuel

Fuel Delivery System Tests

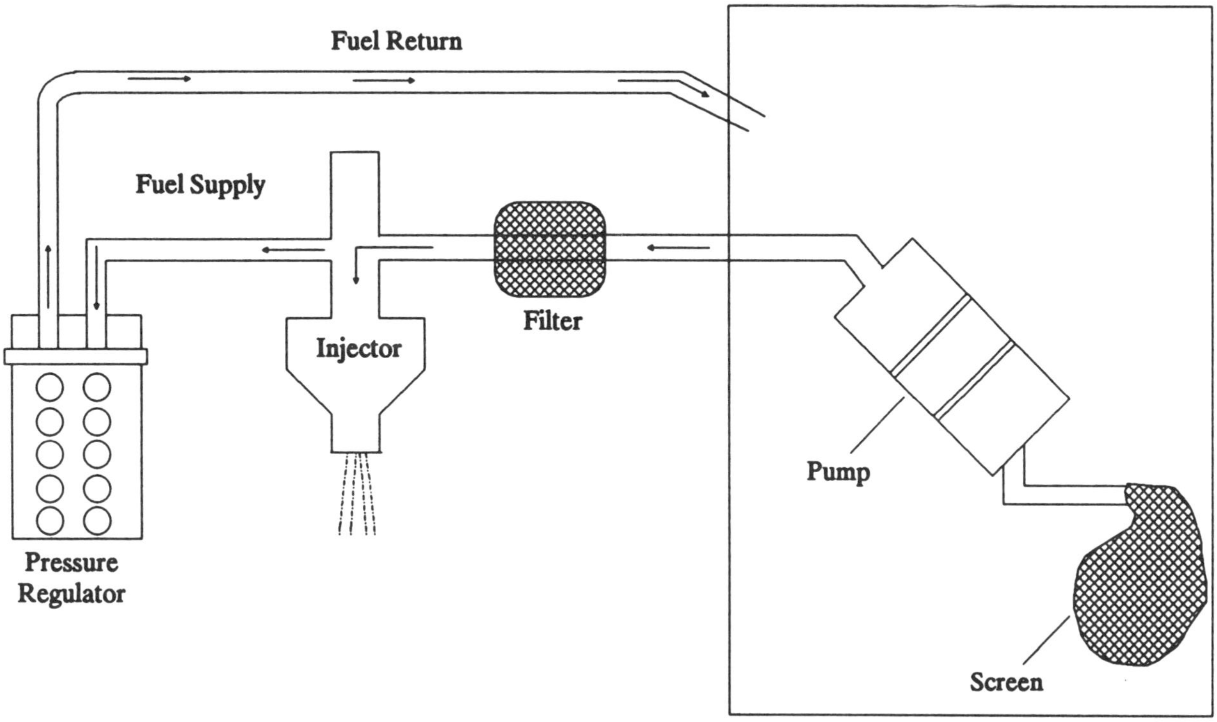

Note: Test pressure anywhere between pump and regulator on the fuel supply side. Most PFI engines will have a test port; Most TBI engines will not.

Testing the Fuel Delivery System is quite simple. Follow the procedures outlined on the next page.

Fuel Delivery System Tests

Cont.

Attach the appropriate fuel pressure gauge to the system. The PFI engines are high pressure and the TBI engines are low pressure. Turn the ignition key ON. Read the STATIC PRESSURE. Compare the reading to the specification listed in Section I. If within specs, the Static Pressure is OK. If the pressure is above the spec, check for a restricted return line. If the return is not restricted, then the fuel pressure regulator is defective. If the pressure is below specs, then run a FUEL FLOW TEST. You must activate the fuel pump for 30 seconds without running the engine. GM EFI engines may have a fuel pump test lead, or the ALDL may have a fuel pump test terminal (Terminal G), or you may have to jumper the fuel pump relay terminals. If the flow is within specs, then the rail pressure is low because the fuel pressure regulator is defective. If the flow is low, check the flow in front of the fuel filter. If the flow rate is good before the filter, then the filter is restricted. If the flow is still below specs, then the pump is defective or the pump filter screen is restricted. Be sure you check the fuel feed line for restrictions before condemning the pump. You can also check the fuel pump DEADHEAD PRESSURE by restricting the return line while activating the fuel pump. Be careful, this procedure can damage the hose inside of the fuel tank if the pressure is kept too high for too long.

Pressure in spec
Static pressure OK

Pressure too high
Restricted return or
Defective Regulator

Pressure too low
Restricted line, filter
or bad pump

Deadhead Presssure –
Three times the spec
pressure – normal

If the STATIC PRESSURE is OK, it does not necessarily verify that the system can deliver enough fuel for all engine running demands. To test the pressure dynamically with the engine running, start the engine. If the pressure regulator is vacuum referenced, which all PFI regulators are, then the pressure should drop when the engine is started. Cycling the throttle should cause the pressure to change up and down as the throttle is changed. If it doesn't, check the vacuum supply tube to the regulator. There should be full manifold vacuum present at the vacuum supply

Fuel Delivery System Tests

Cont.

tube. There should also never be any fuel present in this tube. Rev the engine up to cruise RPM. The fuel pressure should remain within the spec. If it doesn't, run the FUEL FLOW TEST.

Low pressure TBI engines do not use a vacuum referenced regulator. PFI engines do.

PFI Engine
Pressure changes
with vacuum

TBI Engine
Pressure remains
steady

Another fuel delivery system problem can be leakdown. When the engine is shut off, the fuel rail is designed to maintain rail pressure for at least five minutes to facilitate quick restart. To check for leakdown, check the rail pressure after the engine is turned OFF. Observe the rail pressure after five minutes. If the regulator is not leaking, no fuel should have returned to the tank through the regulator. If the fuel pump check valve is not leaking, then no fuel can return to the tank through the feed line. The injector(s) are electrically OFF, and should not allow the fuel to bleed off. If the pressure has dropped, there is leakdown. Restrict the return, and retest. If the leakdown stops, then the regulator is leaking. If the leakdown still occurs, then restrict the feed line. If this corrects the problem, then the leakdown is caused by a bad fuel pump checkvalve.

If restricting both lines still does not stop the leakdown, then the injector(s) or fuel rail is leaking. A quick test to check for leaky injectors is too run the engine up to a high RPM and turn the ignition key OFF. The engine should shut down. If the engine continues to run on, then there is fuel available to the intake system. Non-leaking injectors should normally prevent such an occurrence.

Leakdown

Initial Reading

Restricted Return – Leakdown Stops – Bad Regulator
Restricted Supply – Leakdown Stops – Bad Fuel Pump (Check Valve)
Both Restricted – Leakdown Occurs – Leaky Injectors

Fuel Quality Testing

EFI engines must have good, clean, quality gasoline in order to operate properly. The computer pulse width calculations are based on the fuel quality being constant. The computer calculations will not properly adjust for alcohol-diluted fuel. When the technician encounters a driveability problem on an EFI engine, he should not assume that the fuel quality is not the problem. A good EFI technician will always test the fuel system pressures and flow before being concerned with computer parameters. When a flow test is run, the technician will have captured a good sample of the vehicle's fuel from the bottom of the fuel tank. At that time, it would be a very good idea to test the fuel sample for contamination.

Visual inspection of the fuel sample will reveal whether any visible particulate matter is present. If so, don't forget that the fuel travels through a pick-up screen in the tank, then through a fuel filter, then through screens at each injector. Therefore, any particulate accumulation can be deposited throughout the system and a simple fuel change or filter change may not clear up the whole problem.

Large amounts of water can be seen if present. However, even small amounts of moisture can cause driveability problems. Use water-detecting paste to detect any water contamination.

Alcohol is a common cause of driveability problems or complaints. Use the appropriate tools and procedures to measure the concentration level of any alcohol contained in the fuel.

Reid vapor pressure is the term used to describe the vaporization point of gasoline. If the fuel's Reid vapor pressure point is not correct, engine stalls or hesitations related to vapor-lock problems will show up.

The technician can purchase fuel sample analysis kits which equip him to perform the tests mentioned above. If you ignore the vital factor of fuel quality, you will repeatedly be stung with the embarrassment of misdiagnosis.

Injector Cleaning

The use of fuel injection systems on today's vehicles has produced some great driveability and fuel economy benefits. Fuel injection is a great fuel management system, whereas, carburetion was a great fuel mismanagement system. The major problem associated with fuel injection engines is the fact that the orifices used on fuel injectors are very small and can become plugged or restricted very easily. Particulate matter can plug the injectors instantly or can be dissolved with chemicals. Some cannot. A common substance which often plugs up fuel injectors is a substance called olefin. This is a waxlike substance which is present in almost all gasoline sold in this country. On some vehicles, olefin will build up at a very rapid rate. On some engines, the time necessary for the accumulations to cause problems is much longer. Ten years of domestic EFI engine service in this country has shown that the olefin build-up can and does cause performance problems. When this happens, the olefins can, in most cases, be cleaned out of the injectors while the injectors are still mounted on the engine. This procedure is much less expensive to the vehicle owner and is the preferred procedure to use.

Using the cleaning system of your choice, run a good olefin dissolving solvent through the fuel injectors for approximately fifteen minutes of engine running time. Make sure you run the solvent through the injectors, not through the fuel return line and into the gas tank. This is the key to success. Restrict the return line while performing the cleaning procedure. This will prevent the solvent from heading for the gas tank, provided the fuel pump check valve is okay and doesn't allow the solvent to return to the tank through the fuel supply line.

RAIL ISOLATION METHOD

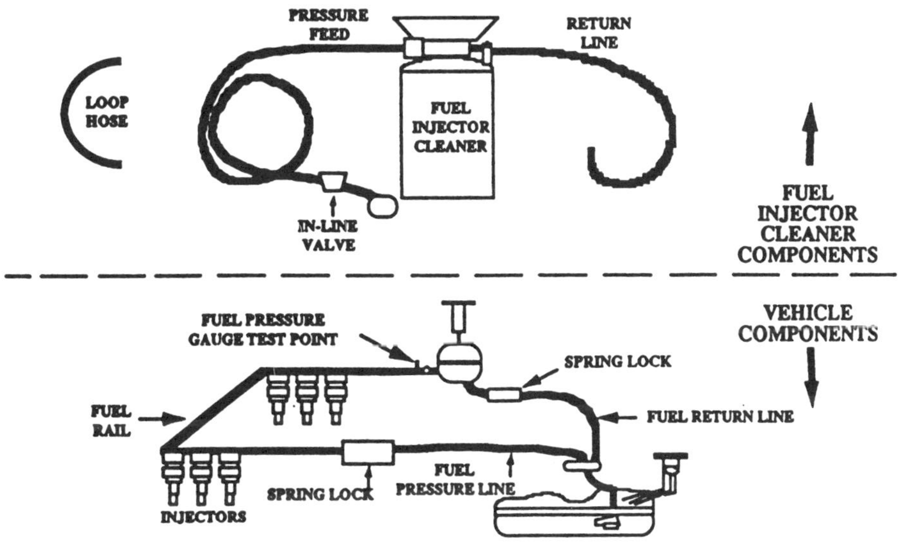

Injector Cleaning

Cont.

Be careful. Plumb into the fuel system carefully. Avoid leaks by using the proper fittings and seals. Make sure the integrity of the fuel delivery system is secure before you return the vehicle to the customer.

Select a cleaning chemical that is safe to the system and to any engine sensors. Preferably, use a solvent which will also dissolve carbon accumulations in the intake manifold and valve system.

Intake Plenum Cleaning

Anytime you encounter rough idle, stalling, or hesitation on an electronic fuel-injected engine, you should check for carbon or gum deposits in the intake plenum assembly. Remove the intake ducts and carefully inspect the throttle body bore and throttle plate. If you see a carbon buildup here, you can bet that the same carbon buildup has probably accumulated in the intake manifold and around the valves.

Use a decarbonization solvent to carefully clean the throttle body bore and the throttle plate. Clean the IAC air passages thoroughly. On some engines, it will be necessary to remove the intake manifold in order to get to the valves and the intake runners properly.

INTAKE PLENUM CLEANING

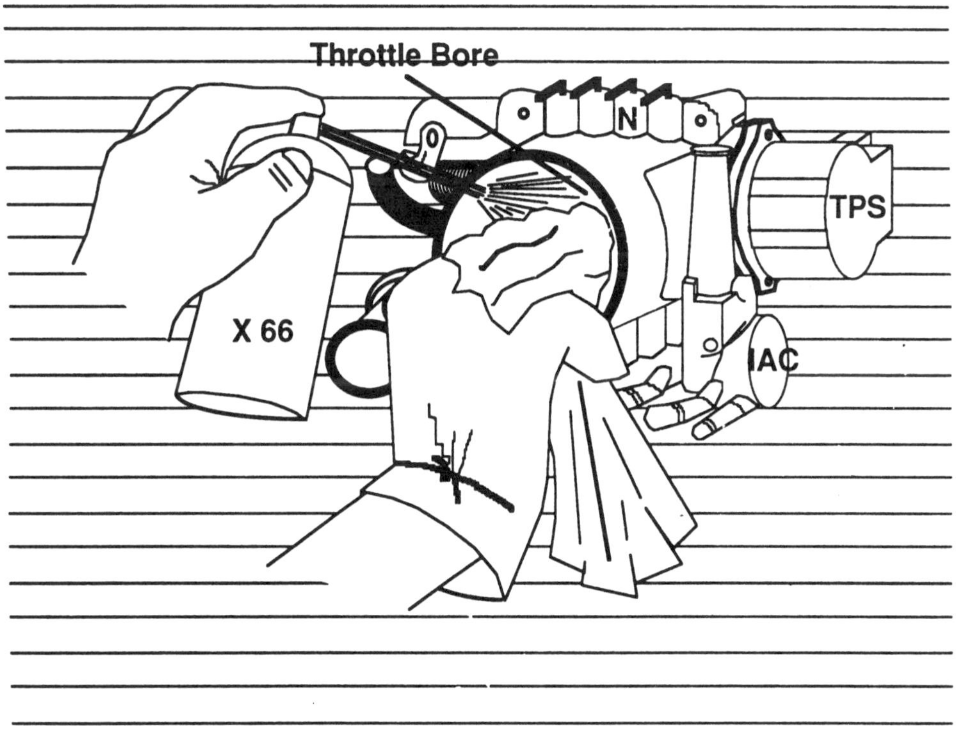

Injector Balance Test

The purpose of running an injector balance test is to determine which injector(s) are restricted, inoperative or delivering fuel differently than its companions. Replacing a complete set of injectors can be quite expensive. The basic tools needed are:

a. Accurate pressure gauge with pressure relief.
b. Injector pulser with time control.
c. Necessary injector connection adaptors.
d. Safe receptacle for catching and disposing of any fuel released.

1. Attach the pressure gauge to the fuel delivery rail on the supply side. Make sure connections are safe and leakproof.

2. Attach the injector pulser to the first injector to be tested.

3. Turn the ignition key to the on position to prime or build up the static rail pressure. Note the reading.

4. Activate the pulser for the precisely-timed firing pulses.

5. Note and record the new static rail pressure after the injector has been pulsed.

6. Repeat this procedure for all of the engine's injectors.

7. Compare the two pressure readings and compute the pressure drop for each injector. Compare the pressure drops of each injector to each other. Any variation in pressure drops will indicate an uneven fuel delivery rate between the injectors.

Example:

Injector	1	2	3	4	5	6
Initial Pressure	40	40	40	40	40	40
Second Pressure	30	30	35	30	20	30
Pressure Drop	10	10	5	10	20	10
Possible Problem	OK	OK	Restriction	OK	Leak	OK

Injector Balance Test

Cont.

This test is best done on a cool engine. The heat of the engine will sometimes cause the fuel in the rail to expand which will cause a pressure variation not caused by the actual delivery rate of the injectors. On some engines, it will be necessary to remove the intake plenum to gain access to the injector connections. This procedure might require a new plenum gasket or seal. In most cases, the removal of the plenum is not a major task. It just looks imposing and difficult to the first timer.

Some engine analyzers have the ability to perform a running engine power balance test and monitor the fuel changes by measuring the HC change as each cylinder is disabled. This procedure is very tricky on EFI engines because the computer will compensate for the cylinder changes by adjusting timing, fuel and idle. Also, the catalytic converter will change its combustion rate as more unburned fuel accelerates the combustion conversion rate.

Therefore, the electromechanical injector power balance test described here is by and large the most reliable way to verify actual per-cylinder fuel delivery. It involves more labor and costs more, but why test at all if you can't test accurately?

Some injector power balance testers are equipped to measure the fuel flow instead of the pressure drop changes. This involves flow vials or flow transducers which are more costly. Whatever tools or testers you decide to use, make sure the readings are accurate.

This test will help you to come to a more conclusive diagnosis on engines with cylinder misfire or rough or uneven running conditions. Fuel injection is great, but injectors can fail too.

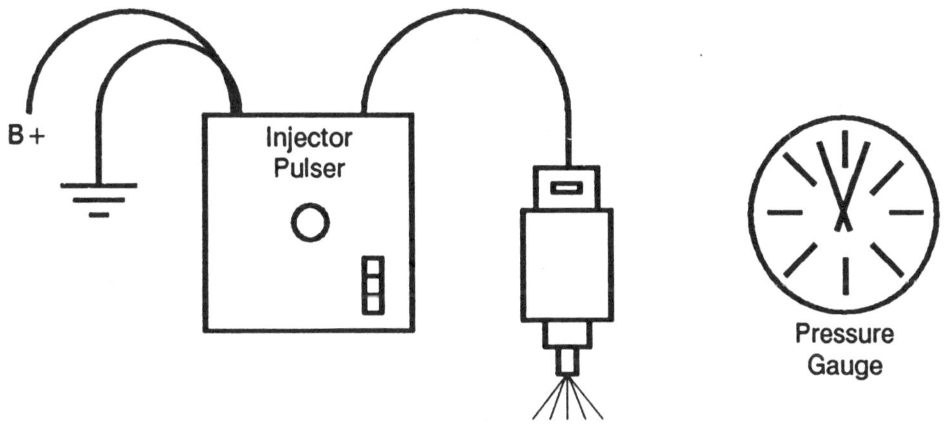

Minimum Idle Tests

On GM EFI engines, the ECM is in control of the engine curb idle speed. The ECM is wired to a control device, usually an idle air control stepper motor, which the ECM uses to control the amount of air bypassed around the throttle plate. This is the basic philosophy used by most GM EFI engine controllers to control the engine curb idle speed as loads are added or removed. A few GM EFI engines use a two-wire DC reversible motor called an ISC. This device actually opens and closes the throttle plate slightly to compensate for loads applied or removed. The ECM can also use the IAC air bypass capability to aid in starting. The actual control and testing of the IAC and the ECM IAC control circuits is more fully discussed in Section III under Idle Air Control.

The Minimum Idle Speed is the term used to describe what the engine idle speed would be if all computer assisted idle modification was removed. The Minimum Idle Speed, therefore, is actually the engine idle speed which is the result of the throttle plate angle at full rest with all bypassed air stopped. Under normal operating conditions, the engine should not be required to operate at this minimum idle speed, but, during certain maneuvers the ECM may not have sufficient time to send the necessary compensation commands to the IAC or ISC to catch and compensate for all engine loads. This means that the minimum idle speed should be sufficient to prevent engine stalling in the event that sudden maneuvers catch the ECM in a compromising position.

Many GM EFI engines have been guilty of the "Phantom Stall". The symptom reminds you of the old carburetor fuel bowl level problem. Many unnecessary parts have been replaced such as MAFs, ECMs, etc., trying to cure this stall problem. The properly trained EFI technician will always check, and adjust if necessary, the engine minimum idle speed. How? It's easy if you follow these steps:

1. Since GM EFI engines use several different control devices to control curb idle speed, the technician must first identify the type of control device used on the vehicle he is testing. The ECM zero references the IAC differently on some models, so choose the steps which work best on the model being tested.

2. Force the engine to its minimum idle speed. Shut off all bypassed air. Use Idle Air Bypass Plugs for the TBI engines. On the PFI engines, you must use a stepper motor driver/tester such as the #222 made by Concept Technology,

Minimum Idle Tests

Cont.

Inc. This device will allow you to force the IAC to close off the idle air passage. On the EFI engines which use an ISC, you can fully retract the plunger from the throttle linkage using a motor driver such as the #221 made by Concept Technology. To prevent the ECM from attempting to readjust the engine idle speed, you must disconnect the idle speed device used.

Idle Air Control (IAC) Motor Tester

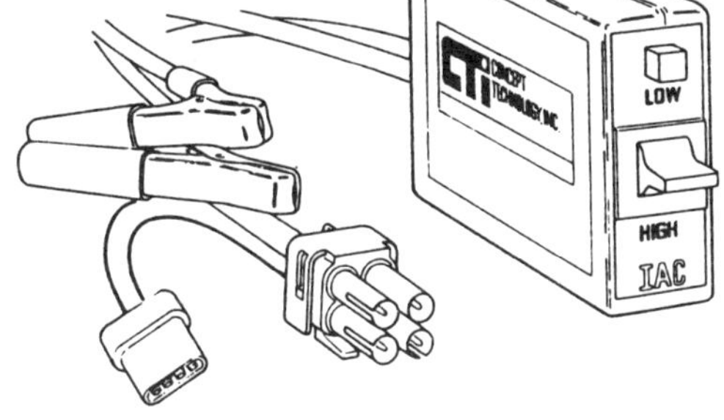

Idle Speed Control (ISC) Motor Test

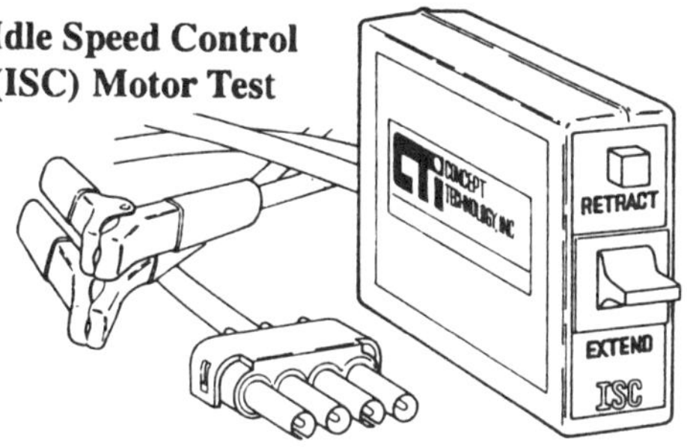

3. Start the warmed engine and monitor the engine base or minimum idle speed. Compare it to the specs listed in Section I. If necessary, adjust the throttle adjustment screw to obtain the proper minimum idle speed. GM throttle bodies always have a throttle adjustment screw. It is usually covered with a tamper resistant plug or it may be covered with a tamper resistant material such as Loctite. You can adjust this throttle plate minimum angle if youneed to. The illustrations of the GM throttle bodies located in Section I will help you in locating the throttle adjustment screw.

4. After you have adjusted the minimum idle speed to the correct setting, then shut the engine OFF. Reconnect the idle speed control device and restart the engine. You will probably have to road test the vehicle for several miles to reteach the IAC zero reference position. This can sometimes be accomplished in the stall by merely idling the engine and shutting the key OFF, but some models require an actual road test above 35 MPH in order for the ECM to reset the IAC zero reference.

Minimum Idle Tests

Cont.

5. While road testing the vehicle, apply and remove various engine loads and note whether or not the ECM can send the correct commands to the IAC or ISC to compensate properly. NEVER adjust the minimum idle speed too high or the ECM may keep setting a Code 35 because the engine idle speed will be above the desired idle speed window which the ECM monitors during normal engine operation. If you have to set the idle speed too high in order to keep the engine from stalling, you are probably just covering up another problem which should be corrected. On engines which use the ISC, be sure to check the NOSE switch. The ECM must have this important input signal to properly control the ISC.

Using A Logic Probe

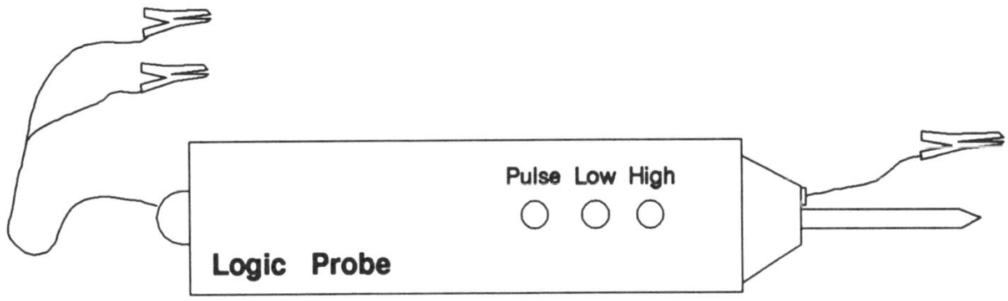

The digital logic probe is a device designed for use by electronic technicians to test a logic circuit. It can be used on vehicles with on-board computer circuits as an easy-to-use circuit voltage indicator. Since the light bulbs used are LEDs, the current draw is minimal and this makes the tester ideal for computer circuit testing. You can still use a standard automotive test light on high current circuits. But you should not use it on low current circuits. There are three LEDs on the logic probe: one red, one green and one yellow.

Think of the red LED as being the standard test light bulb. When the probe is touching a source of voltage, the red LED will illuminate. The difference here is that is only takes 2.2V to turn on the red LED. The green LED is able to tell you something that a standard test light can't. If the green LED illuminates, you know that the voltage at the test point has to be below 0.8V. Hence, the green LED is an ideal indicator with which to test a ground. If the ground is open, the load on the circuit line will raise to pass the voltage through to the ground. Then the red LED would illuminate.

The yellow LED is also a unique addition to your testing information. If the logic probe tip is touching a test point where a DC digital pulse is occurring, the yellow LED will be illuminated. Today's vehicle computers receive and emit a lot of DC digital pulses.

When a DC digital pulse is detected, the red and green LEDs will alternately blink along with the yellow LED. If the red LED is brightest, then the pulse signal is at the high level most of the time. If the green LED is brightest, then the pulse signal is at the low level most of the time. This feature will tell you if you are detecting a square wave or a rectangular wave pulse. If the green and red LEDs are both illuminated at the same level, then you know that the pulse is squared-on 50% of the time and off 50% of the time. The only trick here is that the illumination level of the green and red LEDs is keyed to the power supply voltage of the tester. If you have the probe power leads attached to a 12V source, then that sets the set level of the LED's at 12V. So, if you are probing a DC pulse of 0.0V to 5.0V, then the green LED will appear brightest. If your pulse is 0.0V to 12.0V, then the set

Using A Logic Probe

Cont.

level is at 50% of 12.0V and the green and red LEDs will flash equally if you are detecting a true square wave.

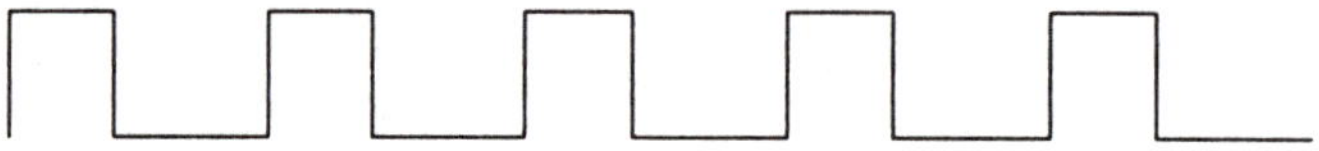

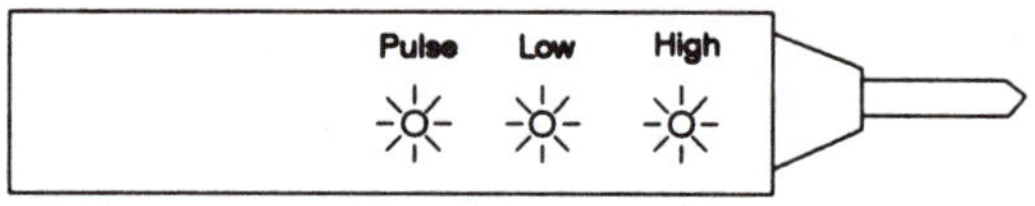

If you are detecting a true square wave pulse, the yellow LED will be blinking, the red and green LEDs will be alternately blinking, provided your power supply source is at the same level as the digital pulse you are detecting.

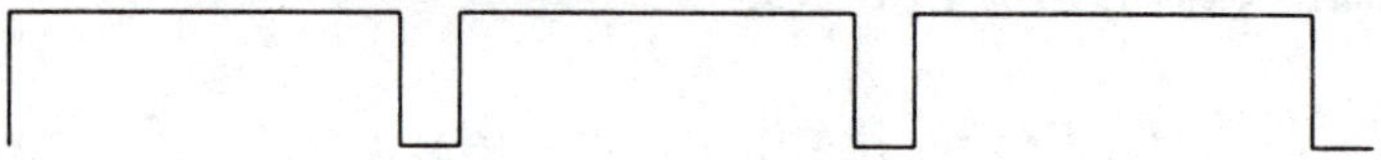

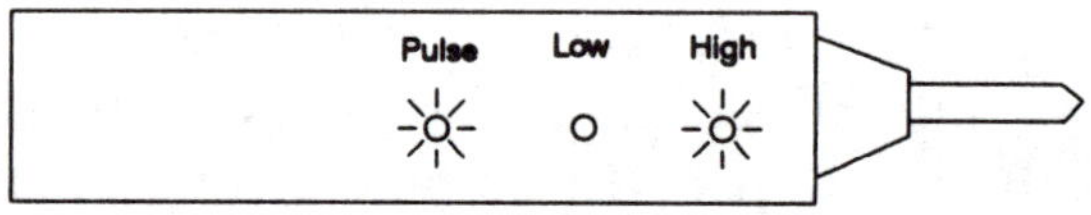

If the pulse is rectangular in shape, the red LED will be brightest if the pulse is high more than it is low.

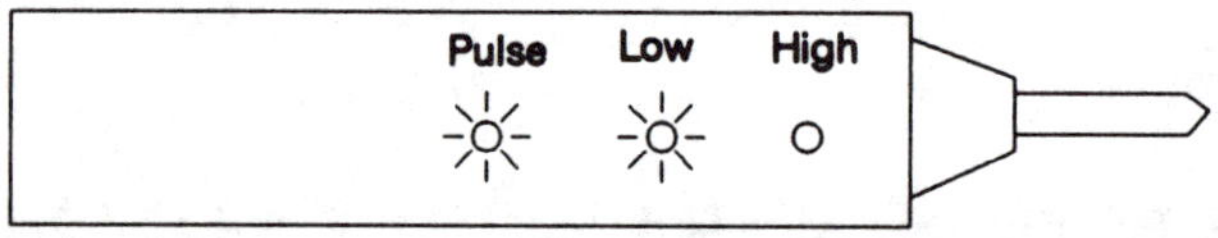

Or the green LED will be brightest if the pulse is low more than it is high.

In summary, using a logic probe is like having three different test lights; one for detecting voltage, one for detecting no voltage or grounds, and one for detecting a pulse signal. Use the probe to verify quickly the presence or absence of any of these circuit conditions.

Using an O-Scope

When computerized engine controls began to appear on domestic vehicles in the early eighties, automotive technicians began to realize that the standard automotive voltmeters in common use were not properly designed to test most sensor and many actuator circuits used on these new microprocessor controlled systems. The tool truckers arrived with the "new" digital voltmeters and quickly convinced many technicians that only digital voltmeters could be used for testing circuits on these new systems.

The great thing about digital voltmeters is that they are usually high impedance. This means that they won't pull much current from the circuit being tested. That's good! What's wrong with these digital voltmeters is the fact that they read constant voltages best. Changing voltages are hard to monitor with a digital voltmeter. Momentary peaks or valleys may or may not be "caught" because the speed at which the digital display is refreshed has to be slowed down sufficiently to give the appearance of readable digits.

The O-scope is a voltmeter which will read voltage, read the polarity of the voltage and informs the user of voltage present over a period of time. The O-scope is the ideal voltmeter for reading voltages on circuits where the voltage is changing.

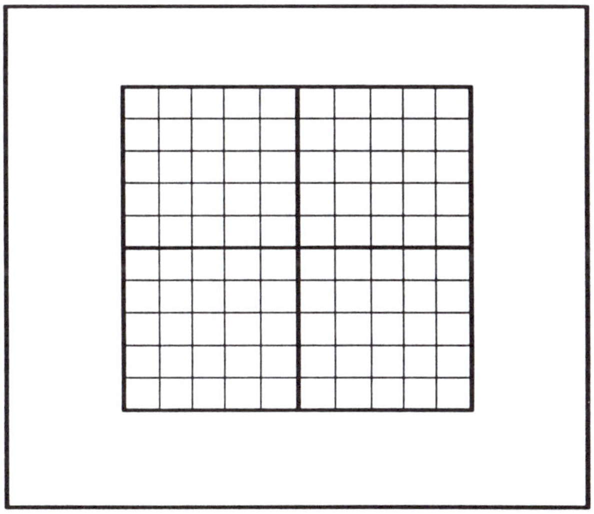

The waveforms displayed on the CRT screen can be measured for voltage strength, voltage polarity and time. The up-down length represents the amount of voltage. The position above or below the zero reference line determines the polarity of the waveform. The length of the signal across the screen determines the time in which the signal occurred. Throughout this manual, there are samples of scope patterns for various signals. These are called signal signatures. If you know what a signal signature should look like, then you can detect problems in a circuit based upon variations in the signal signature.

Technical Service Bulletins

1982 Camaro, Celebrity and Citation

Condition: Intermittent Engine Roughness

Cause: Sudden voltage spike when accessories are operated

Diagnosis & Repair: If this condition exists, the operation of the accessories may cause the engine to run rough due to the ECM misinterpreting the voltage spike and momentarily interrupting the fuel pulses. To repair, install a .50 mfd. capacitor into the IGN circuit to the ECM. This circuit is the pink wire with a black tracer terminating at ECM pin 16 of the white ECM connector.

All GM vehicles with HEI

Condition: Hard Starting/Rough Running

Cause: Carbon tracking on the distributor rotor

Diagnosis & Repair: Any time the condition exists, visually inspect the rotor on the top and bottom for any evidence of carbon tracking. Always check the seal between the ignition coil and the distributor cap, making sure it is greased properly to prevent moisture from entering. Make sure the distributor cap carbon is properly installed. A wide gap between the carbon brush and the rotor tab can cause rotor burn and/or cap destruction.

Technical Service Bulletins (Cont.)

All 4.1L V8 engines

Condition: Poor Engine Performance

Cause: Engine vibrations causing the ESC system to improperly retard the spark

Diagnosis & Repair: Check the engine timing with an advance control timing light. If the spark retards when the engine is revved up in the stall, the ESC controller is retarding the spark due to knock sensor noise detection. Listen to the engine with a stethoscope to determine if there are any abnormal internal engine sounds such as crank knock, etc. If there are none, disconnect the detonation sensor and road test the vehicle. If the performance is restored, replace the ESC controller to stop the improper spark retardation.

All GM Vehicles with EFI

Condition: Rough Idle, Improper Idle Speeds or Surging

Cause: Inoperative or disconnected VSS

Diagnosis & Repair: The VSS is a mandatory input for the ECM to operate three different controlled systems – the Idle Air Control, the TCC and the Canister Purge system. If the VSS signal is not received because the sensor does not work or has been disconnected, the ECM cannot do its job in controlling these three systems. Check the VSS operation and determine the cause(s) of the missing signal.

Technical Service Bulletins (Cont.)

1981-1983 GM EFI Engines with MAP and BARO Sensors

Condition: MAP or BARO inaccuracies

Cause: Bendix (round) sensors

Diagnosis and Repair: When any Bendix type sensors are found in the field, GM has recommended that they be replaced with the Delco type. Be sure they are properly mounted to prevent damage.

All TBI Equipped Engines

Condition: Hesitation

Cause: MAP vacuum supply restriction

Diagnosis & Repair: Tee in a vacuum gauge to monitor the vacuum supply to the MAP sensor. Attach another vacuum gauge to a different full manifold vacuum source. As the throttle is opened and closed, verify that the vacuum signal to the MAP responds to engine vacuum changes as rapidly as the other tested vacuum port. If not, inspect the vacuum source port which supplies vacuum to the MAP sensor. Clean if necessary.

Technical Service Bulletins (Cont.)

1982-1985 Phoenix or 6000 with 2.5L engine

Condition: Intermittent "Check Engine" light

Cause: Shorted wiring harness at the alternator bracket

Diagnosis & Repair: Inspect the wiring harness where it passes near the alternator bracket. Repair if necessary.

1985 2.8L MPFI Engines

Condition: Fuel fouled spark plugs

Cause: Incorrect fuel injector spray pattern

Diagnosis & Repair: If the plugs are fouled, replace the injectors with the new design injectors. GM kit #17111297 includes six new style injectors.

1982-1985 GM EFI Vehicles

Condition: False Code 42

Cause: Electrical feedback from the generator light circuit

Diagnosis & Repair: Always prove that the code is false by following proper diagnostic procedures for a code 42. If the code is false, install a diode jumper assembly, part #12047479 in the generator harness.

Technical Service Bulletins (Cont.)

All GM EFI Engines

Condition: Inoperative Oxygen Sensors

Cause: Silicone contamination

Diagnosis & Repair: Use only sealants which are approved for use on oxygen sensor equipped engines. GM approves two types of sealants – GMS Type A-Clear and GMS Type B-Black.

1985 5.0L TPI Engines

Condition: False Code 34

Cause: PROM error

Diagnosis & Repair: If diagnosis confirms the code is false, install a new PROM, part #127741 w/AC and #1227742 w/o AC.

1980-1984 GM Engines

Condition: Code 14 or 15

Cause: Poor Coolant Temperature Sensor and harness connections

Diagnosis & Repair: If the engine has the early style Weatherpack sensor and harness, replace with kit #10045847.

Technical Service Bulletins (Cont.)

1984-1990 3.0L Engines with CCCI

Condition: Hard Start/No Start

Cause: Improperly grounded ignition module

Diagnosis & Repair: Make sure the module is securely bolted to the mounting bracket. Install an additional ground wire if necessary to ensure good ground contact.

1985-1987 2.8L V6 MPFI Engines

Condition: Clogged injectors

Cause: Pintle design

Diagnosis & Repair: Replace the original Bosch pintle style injectors with the newer Multec ball seat style injectors.

1985-1987 2.8L Engines

Condition: Intermittent Engine Stall

Cause: Residue accumulation inside the throttle bore behind the throttle plates

Diagnosis & Repair: Inspect the throttle bore for residue. If residue exists, clean the throttle bore completely.

Technical Service Bulletins (Cont.)

1987-1990 GM Engines with DIS

Condition:	Poor Engine Performance
Cause:	DIS Coil Secondary Windings
Diagnosis & Repair:	Measure the coil secondary windings resistance both hot and cold. The resistance should be between 5000 and 7000 ohms. This spec does not apply to Type I coils.

1986-1987 2.0L TBI Engines

Condition:	Faulty Injector
Cause:	Shorted injector or weak ECM driver
Diagnosis & Repair:	Measure the injector coil resistance. If less than 1.5 ohms, the cause may be a longer than normal injector ON time. Check for a grounded ECM-to-injector harness. Inspect the ECM label. If it is a round sticker with a silver background and blue letters, the ECM may develop an internal ground. If this is found, replace both the injector and the ECM.

Technical Service Bulletins (Cont.)

1982-1990 GM EFI Engines

Condition: Revised ECM Replacement Procedures

Cause: QUAD Drivers

Diagnosis & Repair: Beginning in 1982, most GM ECMs are equipped with Integrated Circuits in place of separate transistors for various controlled components. The QUAD drivers or QDRs are capable of controlling four different driven circuits. The QDR can lose one output or up to four outputs. This can become confusing to the technician who does not realize that this structure can exist. Before replacing an ECM, determine that all the driven loads are not shorted to ground by using an ohmmeter to verify the proper resistance of the coil. You can also use a tool such as Borroughs BT-8405 to actually measure the current draw of any given device. The current draw should not exceed 750 milliamps for any driven circuit. The QDRs will not handle current in excess of that amount.

1987-1988 Pontiac with 2.8L Engine

Condition: Unstable idle, Stalling and/or Code 35

Cause: Vehicle operator starting the engine with foot on pedal

Diagnosis & Repair: Interrogate vehicle operator to ensure that he/she understands that the start procedure requires no pedal movement.

Technical Service Bulletins (Cont.)

1987-1990 GM Trucks with 7.4L Engine

Condition: Engine Surge

Cause: Fuel vapors in MAP sensor

Diagnosis & Repair: Inspect MAP sensor hose for fuel vapors or residue. Install with the vacuum connection facing downward.

1982-1990 GM EFI Engines

Conditon: Poor Idle, No Start, Code 44

Cause: Defective PCV Grommet

Diagnosis & Repair: Inspect PCV valve grommet for looseness or air leaks. Watch oxygen sensor voltage when jiggling the PCV valve. If the reading was low and rises when the valve is moved, there is an air leak. Replace the grommet.

1984-1990 GM EFI Vehicles

Condition: Repeated Fuel Pump Failures

Cause: Transfer of contaminated strainer

Diagnosis & Repair: Inspect fuel pump pickup strainer for inside or outside restrictions. If restricted, replace the strainer.

Technical Service Bulletins (Cont.)

1988-1990 Pontiac 2.8L Engine

Condition: Cold Start Stall

Cause: Accumulated residue in throttle body bore

Diagnosis & Repair: Inspect throttle body bore carefully for residue buildup. Clean if necessary.

1988-1990 Pontiac 2.8L Engine

Condition: Cold Hesitation

Cause: Lean ECM calibration

Diagnosis & Repair: Follow normal diagnostic procedures for hesitation. Check & adjust TPS to .62-.67V. Ensure there are no leaks to MAP sensor. Check and adjust minimum idle speed. Verify fuel pressure, flow and quality. If all is OK, replace MEM-CAL unit with manufacturer's correct replacement.

Technical Service Bulletins (Cont.)

1988-1989 C/K PU's w/4.3L Engine

Condition: Rough Idle & Converter Odor

Cause: Oxygen Sensor location and ECM calibration

Diagnosis & Repair: Install new "Y" pipe, #15620225 and install oxygen sensor in new "Y" pipe. Install a plug #10017198 in old oxygen sensor hole. Install new PROM according to manufacturer's latest recommendation.

All GM Vehicles

Condition: Reconstituted Coolant

Cause: Service requirement

Diagnosis & Repair: GM does not recommend the use of reconstituted coolant in its vehicles. During normal use, ethylene glycol becomes oxidized, producing a degraded and deteriorated substance. Engine coolant can also become contaminated during draining, handling and storage. For these and other reasons, GM does not believe that reconstituted coolant can meet its specifications.

Technical Service Bulletins (Cont.)

1988-1990 Pontiac Lemans w/1.6L Engine

Condition: Incorrect Timing

Cause: Using wrong mag timing offset

Diagnosis & Repair: The use of the standard 9 1/2 degree timing offset is incorrect for this engine. Refer to the GM service manual for the correct offset.

1988-1990 GM 2.8L & 3.1L Engines

Condition: IAC Zero Reference Error

Cause: New GM ECM "Auto Zero" IAC System

Diagnosis & Repair: If TPS value is read as a percentage on a scan tool, the closed throttle reading should be 0-1%. If above that number, the ECM may have an invalid closed throttle reference for TPS. Check to see if the actual TPS signal voltage is between .29-.98V. Disconnect the TPS harness. Cycle the ignition key . This should cause the ECM to reset the TPS closed throttle reading to zero percent.

Technical Service Bulletins (Cont.)

1987-1990 C/K PU's w/ 7.4L Engine

Condition: False Code 32

Cause: PROM error

Diagnosis & Repair: Verify that the Code 32 is false. If so, then replace PROM with #16145943 for 3.42 and 3.73 axles, or #16145944 for 4.10 and 4.56 axles.

1988-1989 Bonnevilles w/3800 Engine

Condition: Engine Run-On

Cause: Misrouted wire near alternator

Diagnosis & Repair: Check and if necessary adjust minimum idle speed. Check the MAF/TPS wiring branch at the back of the alternator. If pink/black wire is chafed against alternator B+, the back-feed can keep ignition to ECM "hot".

Technical Service Bulletins (Cont.)

1987-1989 Firenza w/2.0L Engine

Condition: Poor Cold Driveability

Cause: Fuel collecting in EGR passages

Diagnosis & Repair: Inspect for fuel accumulation in EGR passage. If so, then install deflector plate #10112342 to prevent fuel from being ingested.

PROM and MEM-CAL Updates

There have been numerous bulletins issued by GM concerning PROM and MEM-CAL updates to cure driveability problems. The list is so long, it would be too difficult to include in this publication. Also, the updated part numbers for these replacements are revised many times after the update bulletin has been issued. The GM dealers can use the CAMS equipment to access the latest information concerning PROM or MEM-CAL revision levels. It is recommended that when there is a question whether vehicle you are servicing is subject to a PROM update, check with the nearest dealer. Most of the early PROM updates dealt with EGR and cold tip-in problems. Many times, the PROM was replaced and the EGR valve also. Most of the later model PROM or MEM-CAL updates deal with actual driveability corrections achieved through recalibrations rather than hardware changes.

The GM ECMs which have Electronically Erasable PROMs, will eventually be reprogrammed in the field by sending new software instructions into the PROM via the ALDL. No hardware replacement will be necessary. The capability now exists to perform this procedure with special equipment, but so far the accepted procedure is to physically replace the PROM. The information contained in the PROMS and MEM-CALS is considered proprietary and GM has been hesitant to make the technology available to the service field. The EPA apparently doesn't want to enable the field technician to change ECM calibrations for fear that emission controls might suffer.

The number of PROM or MEM-CAL changes that have occurred do not represent a large percentage of GM vehicles. Don't assume that that is the main fix these days. Most engine performance problems can be diagnosed and cured using common logical diagnostic procedures and fixes.

Glossary

A/C Switch
The A/C switch sends a signal to the ECM when the air conditioner is operating so engine rpm may be increased to compensate for added load.

Air Injector Reaction System (AIR)
Air flow from pump is directed into engine exhaust manifold and/or converter to reduce exhaust emissions.

Alternating Current (AC)
An electric current that has polarity constantly changing from positive to negative and back again.

Assembly Line Communication (Data) Link (ALCL/ALDL)
Used to flash the "SERVICE ENGINE SOON" or "CHECK ENGINE" light if there are trouble codes.

Amperage
The total amount of current (amperes) flowing in a circuit.

Ampere (Amp)
The unit of measurement for the flow of electrons in a circuit.

Analog Signal
An electrical signal that varies in voltage within a given parameter.

Barometric Absolute Pressure Sensor (BARO)
Reads atmospheric pressure.

Battery
A group of two or more cells connected to produce a current by converting chemical energy into electrical energy.

Bias
The term applied to the electrical, mechanical, or magnetic force or voltage applied to a device to establish the reference point for operation. In a vacuum tube, the bias is the voltage measure between the cathode and the grid.

Binary
A numbering system using a base number or radix of 2. In the binary system, the only two digits are 1 and 0.

Glossary (Cont.)

Binary Counter A counter which produces an output pulse for each two input pulses.

Bit The smallest unit of information recognized by a computer.

Block Learn The portion of ECM memory that adjusts the air/fuel ratio on a semi-permanent basis.

Buffer A part or circuit used to reduce interaction between two electronic circuits.

Byte A unit of information consisting of 8 "bits" that the computer reads and processes.

Computer Controlled Coil Ignition (C^3I) Produces the ignition spark without the aid of an ignition distributor.

Calibration Calibration Memory. The memory which contains an engine's specific calibration.

Calibrator (Mem-Cal) An electronic component that plugs into the ECM which can be specifically programmed to meet engine operating requirements for each vehicle model.

Calibration Package (CALPAC) A device used with fuel injection systems to allow fuel delivery in the event of a PROM or ECM malfunction.

Camshaft Position Sensor The camshaft position sensor sends an electrical signal to the ECM to calculate top dead center of the cylinders in order to properly time ignition and fuel delivery.

Capacitor A device for holding or storing an electrical charge.

Car Inertia Weight Class Weight of a car. Used in exhaust emission tests to determine inertia weight settings for the chassis dynamometer. This information is used to calibrate the engine calibration unit (PROM).

Carbon Monoxide (CO) One of the pollutants found in engine exhaust.

Cathode Both the electron-emitting element of a vacuum tube and the negative electrode of a device.

Glossary (Cont.)

Cathode Ray Tube (CRT) A tube in which the electrons can be focused in a small spot on a fluorescent screen at the opposite end of a structure. By varying the position of the pixels, a picture or pattern can be produced on the screen. Examples of CRTs are a television picture tube or an oscilloscope display tube.

Catalytic Converter Three-way exhausted converter containing platinum and palladium to speed up conversion of HC and CO, and rhodium to accelerate conversion of NOx.

Check Engine (CE) Lights when a malfunction occurs in the ECM system.

Circuit The complete path provided for current flow.

Circuit Breaker A device, other than a fuse, for automatically opening or breaking an overloaded circuit.

Clock A pulse generator in a digital computer which produces pulses to synchronize the switching circuit timing with the memory.

Closed Circuit A circuit which is uninterrupted from the current source and back to the current source.

Closed Loop A system is said to be closed loop when the ECM controls fuel delivery based on the oxygen sensor signal.

Closed Loop Carburetor Control (CLCC) Used to describe oxygen sensor to ECM to M/C solenoid circuit operation.

Cold Start Injector/Valve An auxiliary fuel injector/valve which injects additional fuel into the intake manifold during cold engine starting and operation.

Collector In a transistor, the element that collects charge carriers. It corresponds roughly to the plate of a triode tube and is usually the electrode from which the output is taken.

Collector Current The direct current that flows through the collector of a transistor.

Communication The transmission of information from one point, person or piece of equipment to another.

Glossary (Cont.)

Computer	Any device capable of accepting, comparing and integrating information, performing mathematical calculations and then supplying the results of these processes in proper form.
Computer Command Control (CCC)	Has an electronic control module to control air/fuel ratio and emission systems.
Conductance	A measure of the ease with which a conductor allows electrons to flow. In DC circuits, it is the opposite of resistance.
Conduction	The transmission of heat or electricity through or by means of a conductor.
Conductor	Usually a wire or other metallic object made up of atoms whose free electrons are easily dislodged allowing easy electron flow from atom to atom.
Contact	One of the current-carrying parts of a relay, switch or connector that engages or disengages to open or close the associated electrical circuits.
Continuity	A continuous path for the flow of an electrical current.
Controlled Canister Purge (CCP)	ECM-controlled solenoid valve that permits manifold vacuum to purge the evaporative emissions from the charcoal canister.
Coolant Temperature Sensor	Device that senses the engine coolant temperature and passes that information to the electronic control module through a coaxial connector.
Counter	An electronic circuit that produces an output pulse each time a predetermined number of input pulses is received.
CPS	Abbreviation for Cycles Per Second. Replaced by the term Hertz (Hz).
Cubic Inch Displacement (CID)	The amount of air that all engine cylinders displace.
Current	Amperage. The rate of flow of electrons.

Glossary (Cont.)

Cycle	A complete sequence of a wave pattern that recurs at regular intervals; that is, the change of an alternating wave from zero to a positive peak, back to zero; to a negative peak, and back to zero again. The number of cycles which occur in one second is the frequency of the wave.
Data	Information used as a basis for mechanical or electronic computation.
Diagnostic Code	Pair of numbers obtained from flashing "Check Engine" light or Scan Tool. This code can be used to determine area in the system where a malfunction may be located.
Diagnostic "Test" Term	Lead of ALDL connector which is grounded to get a trouble code. On EFI it is grounded with the engine running to enter the field service mode.
Digital Signal	An electrical signal that is either on or off with no in-between.
Diode	An electronic device that restricts current flow to one direction.
Direct Current (DC)	An electrical current which flows in one direction only.
Direct Ignition System (DIS)	Produces the ignition spark without the aid of an ignition distributor. Same as C^3I.
Driver	An electronic device, usually a power transistor, that operates like a switch; that is, it turns something on or off.
Dropping Resistor	A resistor inserted in series with a line to reduce the voltage to the desired value.
DVM (10 Meg.)	Digital voltmeter with 10 million ohms per volt resistance, used for measurement in electronic systemns.
Dwell	The amount of time (recorded on a dwell meter in degrees) that voltage passes through a closed switch; for example, ignition contact points or internal switch in an electronic control module.
Early Fuel Evaporation (EFE)	Method of warming the intake manifold during cold engine operation. Provides efficient air/fuel mixing.

Glossary (Cont.)

Electric Air Control (EAC) Used on the AIR system to direct air flow to air switching valve of air cleaner.

Electric Air Switching Directs air flow to catalytic converter or exhaust ports of the engine.

Electronic Control Module (ECM) A metal-cased box (located in passenger compartment) containing electronic circuitry that electronically controls and monitors air/fuel ratio and emission systems on Computer Command Control, and turns on the "Check Engine/Service Engine Soon" light when a malfunction occurs in the system.

Electronic Fuel Injection (EFI) A fuel injection system which uses a micro-computer to determine and control the amount of fuel required by, and injected into a particular engine. Computer command control system that uses Throttle Body Injection.

Electronic Module Retard (EMR) Controls spark retard.

Electronic Spark Control (ESC) Used to modify spark advance when detonation occurs.

Electronic Spark Timing (EST) ECM-controlled timing of ignition spark.

Electronic Vacuum Regulator Valve (EVRV) Controls EGR vacuum.

EMI (Noise) An unwanted signal interfering with another needed signal; like an electric razor upsets a television picture or driving under high voltage power lines upsets the AM radio in a car.

Energize/ De-Energize When current is passed through a coil such as the M/C solenoid, the metering control armature is pulled into the solenoid (energized). When the voltage to the solenoid is turned off, a spring raises the metering control armature (de-energized).

Engine Calibration Unit (ECU) An electronic component which can be specifically programmed to the design of each vehicle model to control the M/C solenoid. The ECU plugs into the electronic control module (ECM). The ECU is usually called a PROM.

Glossary (Cont.)

Evaporative Emission Control System (EECS) — Used to prevent gasoline vapors in the fuel tank and TBI from entering the atmosphere.

Federal (FED) — Vehicle/Engine available in all states except California.

Frequency — The number of cycles of a periodic phenomenon in a given unit of time, usually per second.

Fuel Injector — A fuel injector is a spring-loaded, electro-magnetic valve which delivers fuel into the intake manifold, in response to electrical signals from the control module.

Fuse — A device containing a soft piece of metal that melts and breaks the circuit when it is overloaded.

Fuse Link — The current-carrying portion of a fuse that melts when the current is greater than a specific amount.

Ground (–) — A reference point from which voltage measurements may be made. Also, a return path for current to flow back to the source.

Hall Effect Pickup Assembly — This device supplies a basic timing signal to the computer. From this signal, the computer can determine engine RPM, when each piston is coming up on its compression stroke or when engine is in the start mode.

Hertz (Hz) — A term meaning cycles per second.

High — A voltage more than ground or zero. For example, the output wire of an oxygen sensor is referred to O^2 high, as compared to the ground, which is called O^2 low. In digital signals, high is on and low is off.

High Energy Ignition (HEI) — System is a distributor that uses an electronic module and pick-up coil in place of contact points and condenser.

High Impedance Voltmeter — Has high opposition to flow of electrical current. Good for reading circuits with low current flow as found in electronic systems. It allows tests to be made without affecting the circuit.

Hg (Mercury) — A calibration material used as a standard for vacuum measurement.

Glossary (Cont.)

Hydrocarbons (HC)	One of the harmful pollutants found in engine exhaust.
Idle Air Bleed Valve	Controls the amount of air let into the idle fuel mixture prior to the mixture entering the carburetor idle system when the M/C solenoid is energized.
Idle Air Control (IAC)	Device in throttle body units controlled by the ECM to regulate idle air (speed).
Idle Load Compensator	Used to control throttle angle during long deceleration such as coasting down a long grade. Extends at wide-open throttle position or to prevent engine stalls at idle.
Idle Speed Control Motor (ISC)	Regulates throttle valve position and is controlled by the ECM.
Integrated Direct Ignition (IDI)	The type of ignition system used on Oldsmobile 2.3L "Quad 4" engines that produces the ignition spark without the aid of a distributor.
Ideal Mixture	The air/fuel ratio which provides the best performance, while maintaining maximum conversion of exhaust emissions, typically 14.7:1.
Impedance	The total opposition a circuit offers to the flow of current. It includes resistance and reactance and is measured in ohms (i.e., 10 megaohms).
Inputs	Information from sources (coolant temperature sensors, exhaust oxygen sensor, etc.) that tells the ECM how the engine is performing.
Integrator	ECM memory that temporarily adjusts air/fuel ratio.
Intermittent	Occurs now and then, not continuously. In electrical circuits, refers to occasional open, short or ground.
Kilometers Per Hour (km/H)	A metric unit measuring distance (1000 meters) in one hour.

Glossary (Cont.)

L4 — Four cylinder in-line engine.

Liter (L) — A metric unit of capacity.

Malfunction — A problem that causes the system to operate incorrectly. Typical malfunctions are wiring harness opens or shorts, failed sensors or PROM failure.

Manifold Air Temperature Sensor (MAT) — A device that measures the temperature of air in the intake manifold.

Manifold Pressure Sensor (MAP) — Reads pressure changes in intake manifold. Also known as manifold absolute pressure sensor.

Manifold Vacuum Sensor — Reads pressure changes in intake manifold in relation to barometric pressure. May be called a differential pressure sensor.

Mass Air Flow (MAF) — The amount of air flowing into an engine measured by mass instead of density.

Memory Capacity Relationships — A 1k memory can store 1000 bytes. Memory capacities are sometimes expressed as the number of "words" of a certain length that it can hold. One "word" contains two "bytes", similarly, one byte equals eight "bits".

Miles Per Hour (MPH) — An English unit measuring distance (5,280 feet) in one hour.

Micro — A prefix meaning one millionth. Abbreviated μ.

Mixture Control (M/C) Solenoid — Device installed in carburetor which regulates the air/fuel ratio.

Mode — A particular state of operation.

Multimeter — A test instrument with suitable switching facilities to measure voltage, current and resistance. Also called a volt-ohm-milliammeter (VOM).

Glossary (Cont.)

Multi-Port Fuel Injection (MFI/MPFI)	Individual injectors for each cylinder are mounted in the intake manifold. The injectors are fired in groups rather than individually.
Newton Meters (Torque)	A metric unit which measures force.
Nitrogen, Oxides of (NOx)	One of the pollutants found in engine exhaust.
Non-Volatile Memory	Memory retained in block learn cells (not affected by turning the ignition on and off).
Normally Closed	State of relay contacts or solenoid plunger when no voltage is applied.
Normally Open	State of relay contacts or solenoid plunger when no voltage is applied.
Open Loop	Describes ECM control of the M/C solenoid without use of oxygen sensor information.
Output	Functions, typically solenoids, that are controlled by the ECM.
Oxygen (O^2) Sensor	Monitors the content of the exhaust system and generates a voltage signal to the ECM.
Park/Neutral (P/N) Switch	Used on vehicles equipped with automatic transmissions, the P/N switch provides a signal to the ECM to indicate that the transmission is in a driving mode and not in Park or Neutral.
Piezoelectric	The property exhibited by some crystals which causes a voltage to be produced when it is subjected to mechanical stress and, conversely, a mechanical stress to be produced when it is subjected to a voltage.
Positive Crankcase Ventilation (PCV)	Prevents fumes in crankcase from passing into atmosphere.
Pressure Regulator	A device to control the pressure of fuel delivered to the fuel injector(s).
Program Memory	The memory that includes the engine operating instructions.

Glossary (Cont.)

Programmable Read Only Memory (PROM) An electronic component that plugs into the ECM which is specifically programmed to meet engine operation requirements for each vehicle model.

Pulse Air Injector Reactor System (PAIR) Pulse air directed into engine to reduce exhaust emissions.

Pulse Width Modulated (PWM) An electrical device operated by a digital signal that is controlled by the time duration the device is turned on and off.

Quad Driver A semiconductor that is capable of operating four separate outputs. Some have digital outputs and some have pulse width modulated outputs.

Random Access Memory (RAM) The memory in the ECM that stores and can recall from various locations as needed, rather than in sequential order.

Read Only Memory (ROM) A computer memory that is preprogrammed and used primarily for information retrieval applications.

Rear Vacuum Brake (RVB) Used to control choke operation during cold engine conditons.

Relay An electromagnetic device using low current to open or close high current circuits or to interrupt an electrical current.

Resistance The ability of a circuit to limit current flow; like a restriction in a water pipe.

Resistor A device used in an electrical circuit to produce work or to lower the voltage in a circuit.

Rest Pressure Fuel pressure maintained within the system after the engine has been shut down.

Revolutions Per Minute (RPM) A measure of rotational speed.

RFI Filter A device located at the in-tank electric fuel pump. The filter prevents electrical static created by the pump from entering the wiring harness where it could affect other electrical/electronic components.

Glossary (Cont.)

Self-Diagnostic Code	The ECM can detect malfunctions in the system. If a malfunction occurs, the ECM turns on the "Check Engine/Service Engine Soon" light. A diagnostic code can be obtained from the ECM through the CE light. This code will indicate the area of the malfunction.
Sensors	Devices which provide the control module, by electrical signal, with engine information as needed to properly control engine function.
Sequential Fuel Injection (SFI)	On selected applications the injectors are pulsed sequentially (one-by-one) in spark plug firing order.
"Service Engine Soon" Light	Lights when a malfunction occurs in the Computer Command Control system.
Square Wave	An essentially square or rectangular-shaped wave. A wave that alternately assumes two fixed values for an equal length of time with a negligible transition time between the two values.
System Pressure	The level of fuel pressure maintained in the system during normal engine operation.
Tachometer (Tach)	A device for indicating speed of rotation.
"Test" Lead Or Terminal	Lead or ALCL/ALDL connector terminal which is grounded to obtain a trouble code.
Thermal Vacuum Switch (TVS)	Used to control vacuum in relationship to engine temperature.
Thermostatic Air Cleaner (Thermac)	Provides preheated air to outer intake manifold to provide better driveability when engine is cold.
Throttle Body Injection (TBI)	Any of several injection systems which have the fuel injector(s) mounted in a centrally located throttle body, as opposed to positioning the injectors close to an intake port.
Throttle Position Sensor (TPS)	Detects movement of the throttle plate when the throttle is opened or closed and provides the ECM or control module with an appropriate voltage signal.

Glossary (Cont.)

Transmission/ Transaxle ConverterClutch (TCC)	ECM-controlled solenoid in the transmission that engages the torque converter clutch.
Tuned Port Injection (TPI)	The Corvette, Camaro and Trans AM PFI systems are referred to as Tuned Port Injection because of the design of their air induction systems.
V6	Six-cylinder engine arranged in a "V".
V8	Eight-cylinder engine arranged in a "V".
Vac Sensor	Abbreviation for differential pressure sensor, which is a vacuum sensor.
Vacuum	Negative pressure, less than atmospheric pressure.
Vacuum, Manifold	Vacuum source in manifold below throttle plate.
Vacuum, Ported	Vacuum source above throttle plate.
Vehicle Speed Sensor (VSS)	Sensor in speedometer cluster that sends vehicle speed information to the electronic control module.
Volatile Memory	Memory retained in the block learn memory only with the ignition in the ON position.
Volt (V)	Unit of electrical potential difference and electromotive force equal to the difference of potential between two points in a conducting wire carrying a constant current of one ampere when the power dissipated between these two points is equal to one watt and equivalent to the potential difference across a resistance of one ohm when one ampere is flowing through it.
Voltage	The pressure or force pushing the current in a circuit, like pressure in a water pipe.
Wide Open Throttle (WOT)	When the air valves are opened more than 80%, the vehicle is at WOT.
Word	A unit of information consisting of two "bytes"representing the largest data element normally processed by a computer.

Acronyms

A/C	Air Conditioning
AAP	Auxiliary acceleration pump
AC/TVS	Aircleaner bimetal vacuum switch
ACS	Air control solenoid
ADS	Anti-diesel solenoid
AIC	Air intake control valve
AIR–MGMT	Air Management
AIR/CSV	Air injector reaction control valve
AIR/MV	Air management valve
AIR/SSV	Air injector air switch solenoid valve
AIR/VDV	Air injector vacuum delay valve
ALCL	Assembly line communication link
ASS	Air switching solenoid
ASV	Air suction valve
ATS	Altitude trim solenoid
BARO	Barometric pressure sensor
BP/EGR	Back pressure EGR valve
BS	Brake light switch
C/L	Closed loop

Acronyms (Cont.)

C3	Computer command control
C4	Computer controlled catalytic converter
CAM	Cam sensor
CB	Choke breaker
CC	Catalytic converter
CCC	Computer systems
CCIS	Computer controlled coil ignition system
CCP	Control canister purge
CCV	Canister control valve
CDR	Crankcase depression regulator
CEL	Check engine light
CFS	Carburetor feedback system
CHAR/CAN	Charcoal canister
CI	Computer controlled distributorless ignition
CKV	Check valve
CLCC	Closed loop carburetor control
CMH	Cold mixture heater
CO	Choke opener
CPS	Crankshaft position sensor
CPSV	Canister purge solenoid valve
CSFIS	Cold start fuel injector switch

Acronyms (Cont.)

CSTSE	Electric air control/electric air switching valve
CSTTS	Cold start thermal time switch
CSV	Cold start valve
CTOS	Coolant temperature override
CTS	Coolant temperature sensor
DEC	Diesel electronic control system
DEC/V	Deceleration valve
DEFI	Digital electronic fuel injection
DFC	Deceleration fuel injection
DFI	Digital fuel injection
DIS	Distributorless ignition system or Direct ignition system
DPS	Differential pressure switch
DS or KS	Knock sensor
DS/CP/TVS	Distributor spark/canister purge thermal vacuum switch
DS/TVS	Distrubutor spark thermal vacuum switch
DS/VDV	Distributor spark vacuum delay valve
DS/VMV	Distrubutor vacuum modulator valve
DS/VRDV	Distributor vacuum retard delay valve
DS/VRV	Distributor regulator valve
DTCS	Distributor thermal control switch
DTCV	Distributor thermal control valve

Acronyms (Cont.)

DV	Diverter valve
DV/TS	Delay valve trapped spark
DVMV	Distributor vacuum modulator valve
DVSV	Distributor vacuum solenoid valve
EABCV	Electronic air bleed control valve
EAC	Electric air control valve
EAC & EAS	Electric air control valve plus electric air switch valve
EAC & LACN	Electric air control valve & low vacuum air control valve
EBCV	Electronic air bleed control valve
ECES	Electric air control/electric switching valve
ECM	Electronic control module computer
EECS	Fuel evaporative control
EFE	Early fuel evaporator
EFE/TVS	Early fuel evaporator thermal
EFI	Electronic fuel injection
EGR	Exhaust gas recirculator
EGR/CLPS	EGR control lever position sensor
EGR/CM	EGR computer modulated
EGR/CONT	EGR controller
EGR/ESS	EGR engine speed sensor
EGR/QVRV	EGR quick vacuum response valve

Acronyms (Cont.)

EGR/SOL	EGR solenoid
EGR/SR	EGR solenoid & regulator
EGR/TDS	EGR temperature diagnosis switch
EGR/TS	EGR thermal switch
EGR/TVS	EGR thermal vacuum switch
EGR/VSV	EGR vacuum switch valve
EMR	Electronic module retard
EPR	Exhaust pressure regulator
EPR/SOL	Exhaust pressure regulator solenoid
ESC	Electronic spark control
ESCM	Electronic spark control module
ESS	Electronic spark selection Engine speed sensor
EST	Electronic spark timing
EVAP or EEC	Evaporative emissions control system
EVRV	Electronic vacuum regulator valve
FBC	Feedback carburetor
FBVA	Fuel bowl vent actuator
FBVS	Fuel bowl vent solenoid
FIS	Fuel injection solenoid
FITS	Fast idle temperature sensor
FPRA	Fuel pressure regulator

Acronyms (Cont.)

FTS	Fan temperature switch
FVCS	Fuel vapor canister solenoid
HAC	High vacuum air control valve
HAI	Hot air intake diaphragm
HCV	Heat control valve
HEI	High energy ignition
HIC	Hot idle compensator
IAC	Idle air control
IAM	Idle air management
IDI	Integrated direct ignition
ILC	Idle load control
ISC	Idle speed control
ISS	Idle speed solenoid
ISV	Idle speed valve
LAC	Low vacuum air control valve
LAC & EACN	Low vacuum air control plus electric air control valve
LAC & LACN	Low vacuum air control plus low vacuum air control valve
MAF	Mass airflow sensor
MAIR	Managed air injection system
MAP	Manifold pressure sensor

Acronyms (Cont.)

MAT	Manifold temperature sensor
MCS	Mixture control solenoid
MDPS	Manifold differential pressure sensor
MISAR	Microprocessed sensing automatic regulation
MPI	Multipoint fuel injection
MVS	Metering valve system
NDV	Nylon divert valve
O/L	Open loop
O^2	Exhaust gas sensor
PAIR	Pulse air injection reaction system
PCV	Positive crankcase ventilation
PEDES	Pressure operated electric divert and electric switching
PFI	Ported fuel injection
P/N	Park/Neutral switch
PROM	Programmable read only memory chip
PS	Power steering switch
PVS	Ported vacuum switch
PWM	Pulse width modulated solenoid
RAM	Random access memory
RDES	Remote divert/electric switching valve
RVR	Response vacuum reducer

Acronyms (Cont.)

SDV	Standardized diverter valve
SFI	Sequential fuel injection
SSC/SOL	Stepped speed control solenoid
TAC	Thermostatic aircleaner system
TBI	Throttle body injection
TCC	Three-way catalytic converter Torque converter clutch Transmission converter clutch
TOR	Transmission override relay
TPCV	Tank pressure control valve
TPI	Throttle point injection
TPS	Throttle position sensor
TVRV	Transaxle vacuum reducer valve
TVS	Temperature vent switch Thermal vacuum switch
TVSA	Trapped vacuum spark advance
TVSV	Thermostatic vacuum switching valve
VAD	Vacuum advance delay valve
VARV	Altitude vacuum reducer valve
VB/DV	Vacuum break & delay valve
VCV	Vacuum check valve
VDV	Vacuum delay valve
VIS	Vacuum input switch

Acronyms (Cont.)

VR	Vacuum reducer
VRV	Vacuum regulator valve
VS or VAC	Vacuum sensor
VSBA	Vehicle speed buffer amplifier
VSS	Vehicle speed sensor
VSV	Vacuum switching valve
WOT	Wide open throttle switch
ZDV	Zinc divert valve

Notes:

Notes:

Notes: